AF800766

The Saber and Scroll Journal

Also from Westphalia Press
westphaliapress.org

The Idea of the Digital University

Dialogue in the Roman-Greco World

The History of Photography

International or Local Ownership?: Security Sector Development in Post-Independent Kosovo

Lankes, His Woodcut Bookplates

Opportunity and Horatio Alger

The Role of Theory in Policy Analysis

The Little Confectioner

Non-Profit Organizations and Disaster

The Idea of Neoliberalism: The Emperor Has Threadbare Contemporary Clothes

Social Satire and the Modern Novel

Ukraine vs. Russia: Revolution, Democracy and War: Selected Articles and Blogs, 2010-2016

James Martineau and Rebuilding Theology

A Strategy for Implementing the Reconciliation Process

Issues in Maritime Cyber Security

A Different Dimension: Reflections on the History of Transpersonal Thought

Iran: Who Is Really In Charge?

Contracting, Logistics, Reverse Logistics: The Project, Program and Portfolio Approach

Unworkable Conservatism: Small Government, Freemarkets, and Impracticality

Springfield: The Novel

Lariats and Lassos

Ongoing Issues in Georgian Policy and Public Administration

Growing Inequality: Bridging Complex Systems, Population Health and Health Disparities

Designing, Adapting, Strategizing in Online Education

Pacific Hurtgen: The American Army in Northern Luzon, 1945

Natural Gas as an Instrument of Russian State Power

New Frontiers in Criminology

Feeding the Global South

Beijing Express: How to Understand New China

The Rise of the Book Plate: An Exemplative of the Art

The Saber and Scroll Journal
Volume 10, Number 3 • Spring 2022

Westphalia Press
An imprint of Policy Studies Organization
1527 New Hampshire Ave., NW
Washington, D.C. 20036
info@ipsonet.org

ISBN: 978-1-63723-607-9

Cover and interior design by Jeffrey Barnes
jbarnesbook.design

Daniel Gutierrez-Sandoval, Executive Director
PSO and Westphalia Press

Updated material and comments on this edition
can be found at the Westphalia Press website:
www.westphaliapress.org

The Saber and Scroll Journal

Volume 10, Number 3 • Spring 2022

Jeff Ballard, editor-in-chief

Westphalia Press
An imprint of Policy Studies Organization

On the cover: "Welcome Freaks!" proclaims owner and curator of the Pexcho's American Dime Museum, Peter Excho in Augusta, Georgia. If you still have an appetite after viewing the museum's wacky, weird, and sometimes repulsive, artifacts, coffee, soda, and snacks are available for purchase. Photo by Aisha Manus.

APUS

The Saber and Scroll Journal
Volume 10, Number 3 • Spring 2022

Historical Studies Honor Society

cont'd.

PSO

Welcome Letter

Welcome to the Spring 2022 issue of the Saber and Scroll Journal, the first edition in our 9th year in print. My name is Jeff Ballard and effective January 1st, I am the Editor-in-Chief of this humble student and alumni-run venture. In the new year, I have much to report: some of the news is exciting and some rather melancholy.

Beginning with Winter 2021, the Journal and all its past issues were transferred to Scholastica.com (https://saberandscroll.scholasticahq.com). Once indexed by Google, and other search engines, all eight years of the Journal's content will be searchable on the Web. Our authors too, past and present, will be more easily discovered by readers hungry for more of the things they had to say. Now for the sad news.

Every ship has just one captain and because I am on the bridge that means that Lew Taylor has retired. Many thanks for your years of service to the APUS community. I have never known a man who wore so many hats as Lew. If there is a silver lining, it is that Lew has agreed to remain on staff as our Editor-at-Large, with the byline, "Somewhere in the Yucatan." Fair winds and following seas, Mr. Taylor.

What is with all the nautical speak? If you have not guessed, by avocation, I am a naval historian having graduated with a master's degree in military history, with honors, from the American Military University (2015). Though my area of interest is war in the South Pacific War from 1942 to 1943, I'm equally fascinated by the topics that fascinate others.

True, much of what we publish relates to armed conflicts in world history, but we cast our net broadly, to appeal to as many readers as possible. The Saber and Scroll Journal Volume 10 Number 3 (Spring 2022) is no exception.

Filed under "Weird but Wonderful" please enjoy our cover article, Aisha Manus' award-winning *Pexcho's American Dime Museum Review*. From shrunken heads to a vial of flatulence, to Victorian medical (torture?) devices, there is something at Pexcho's to entertain - or revolt - everyone. Even Ms. Manus, a repeat museum patron, was so repulsed by his demonstration she needed someone else to photograph Red the Sword Swallower.

In partnership with the Historical Miniatures Gaming Society – Pacific Southwest chapter, the Journal is proud to introduce AJ Roth, their Rising Historian for 2021. Judges and Journal editors alike were so impressed with this ninth-grade homeschooler that we had to put his manuscript, *This Day I Conquer or Die: The Battle of Bleinheim, 1704*, front and center in this issue. Bravo Zulu (Outstanding!) AJ. Can't wait to read what you write this year.

 doi: 10.18278/sshj.10.3.1

Between these two bookends, we have a fantastic assortment of articles and book reviews: Roman philosophy to natural selection, the Second Punic War to the American Civil War, and Union balloonists go to Hollywood to meet the Sullivan brothers in 1774 The Longest Year of the Revolution.

Finally, I am not able to close this letter without comment on the events unfolding in Europe. On February 28, 2022, the American Historical Association released the statement, "We vigorously support the Ukrainian nation and its people in their resistance to Russian military aggression and the twisted mythology that President Putin has invented to justify his violation of international norms." While the Journal will always endeavor to be apolitical in its content and editorial positions, we cannot be silent on what the AHA calls "… Putin's abuse of history as justification for the attack [on Ukraine]." (Source: https://www.historians.org/news-and-advocacy/historians-condemn-russian-invasion-of-ukraine-(februa ry-2022))

"War is mankind's most tragic and stupid folly; to seek or advise its deliberate provocation is a black crime against all men." Dwight D. Eisenhower, Graduation Exercises at the United States Military Academy at West Point, New York, U.S. (June 3, 1947)

Had the Saber and Scroll Journal Staff been asked to sign the AHA statement, we would have done so most emphatically.

Ex Tenebris Lux
"From Darkness Light"

Jeff Ballard
Editor-in-Chief
Huntington Beach, California
March 8, 2022.

Carta de bienvenida

Bienvenido a la edición de primavera de 2022 de Sabre and Scroll Journal, la primera edición en nuestro noveno año impreso. Mi nombre es Jeff Ballard y, a partir del 1 de enero, soy el editor en jefe de esta humilde empresa dirigida por estudiantes y exalumnos. En el nuevo año, tengo mucho que informar: algunas noticias son emocionantes y otras bastante melancólicas.

A partir del invierno de 2021, la Revista y todos sus números anteriores se transfirieron a Scholastica.com (https://saberandscroll.scholasticahq.com). Una vez indexado por Google y otros motores de búsqueda, los ocho años del contenido de la Revista se podrán buscar en la Web. Nuestros autores, pasados y presentes, también serán descubiertos más fácilmente por los lectores hambrientos de más de las cosas que tenían que decir. Ahora las tristes noticias.

Cada barco tiene un solo capitán y porque estoy en el puente, eso significa que Lew Taylor se ha retirado. Muchas gracias por sus años de servicio a la comunidad APUS. Nunca he conocido a un hombre que usara tantos sombreros como Lew. Si hay un lado positivo, es que Lew ha aceptado permanecer en el personal como nuestro editor general, con el título "En algún lugar de Yucatán". Buenos vientos y buen mar, Sr. Taylor.

¿Qué pasa con toda la jerga náutica? Si no lo ha adivinado, por vocación, soy un historiador naval que se graduó con una maestría en historia militar, con honores, de la Universidad Militar Estadounidense (2015). Aunque mi área de interés es la guerra en la Guerra del Pacífico Sur de 1942 a 1943, estoy igualmente fascinado por los temas que fascinan a otros.

Cierto, mucho de lo que publicamos se relaciona con conflictos armados en la historia mundial, pero lanzamos nuestra red ampliamente, para atraer a tantos lectores como sea posible. The Sabre and Scroll Journal Volumen 10 Número 3 (primavera de 2022) no es una excepción.

Archivado en "Extraño pero maravilloso", disfrute de nuestro artículo de portada, el galardonado Pexcho's American Dime Museum Review de Aisha Manus. Desde cabezas reducidas hasta un vial de flatulencia, pasando por dispositivos médicos (¿de tortura?) victorianos, hay algo en Pexcho's para entretener, o rebelar, a todos. Incluso la Sra. Manus, una patrocinadora habitual del museo, estaba tan repelida por su demostración que necesitaba a alguien más para fotografiar a Red the Sword Swallower.

En asociación con la Sociedad de Juegos de Miniaturas Históricas – Capítulo del Sudoeste del Pacífico, el Journal se enorgullece de presentar a AJ Roth, su Rising Historian para 2021. Tanto los jueces como los editores del Journal quedaron tan impresionados con este estudiante de noveno grado que tuvimos que publicar su manuscrito, *Today I Conquer or Die: The Battle of Bleinheim*, 1704, al frente y al centro de este número. Bravo Zulu (¡Sobresaliente!) AJ. No puedo esperar a leer lo que escribes este año.

Entre estos dos sujetalibros, tenemos una fantástica variedad de artículos y reseñas de libros: de la filosofía romana a la selección natural, de la Segunda Guerra Púnica a la Guerra Civil Estadounidense, y los globos aerostáticos de la Unión van a Hollywood para encontrarse con los hermanos Sullivan en 1774 El año más largo del Revolución.

Finalmente, no puedo cerrar esta carta sin comentar los acontecimientos que se desarrollan en Europa. El 28 de febrero de 2022, la Asociación Histórica Estadounidense emitió la declaración: "Apoyamos enérgicamente a la nación ucraniana y a su pueblo en su resistencia a la agresión militar rusa y la mitología retorcida que el presidente Putin ha inventado para justificar su violación de las normas internacionales". Si bien el Journal siempre se esforzará por ser apolítico en su contenido y posiciones editoriales, no podemos guardar silencio sobre lo que la AHA llama "... el abuso de la historia por parte de Putin como justificación del ataque [contra Ucrania]". (Fuente: https://www.historians.org/news-and-advocacy/historians-condemn-russian-invasion-of-ukraine-(february-2022))

"La guerra es la locura más trágica y estúpida de la humanidad; buscar o aconsejar su provocación deliberada es un crimen negro contra todos los hombres." Dwight D. Eisenhower, Ejercicios de graduación en la Academia Militar de los Estados Unidos en West Point, Nueva York, EE. UU. (3 de junio de 1947)

Si se le hubiera pedido al personal de Saber and Scroll Journal que firmara la declaración de la AHA, lo habríamos hecho de la manera más enfática.

Ex-Tenebris Lux

"De la oscuridad a la luz"

Jeff Ballard

Editor Principal

Huntington Beach, California

8 de marzo de 2022

编者按

欢迎阅读《军刀与卷轴杂志》2022年春季期刊，这是本刊创立以来第9年第1版。我是Jeff　Ballard，自今年1月1日起担任本刊（由学生和校友共同运作）主编。在这新的一年里，我有许多想要倾诉：一些激动人心的消息和一些感伤的消息。

从2021年冬季期刊开始，本刊及其以往发行的内容都转移到Scholastica.com（https://saberandscroll.scholasticahq.com）。谷歌和其他搜索引擎将其列入索引后，可在Web上搜索本刊八年的所有内容。过去和现在为本刊服务的作者也将更容易被其读者所检索。接下来是悲伤的消息。

一艘船只有一个船长，既然我已上任，这便意味着Lew　Taylor离任。在此衷心感谢几年来其对美国公立大学系统（APUS）社区的服务　。我从未见过谁像Lew一样身兼数职。如果还有些许好消息的话，那便是Lew同意作为特约编辑留在我们团队，其署名为“尤卡坦的某处”（Somewhere in the Yucatan）。一路顺风，Taylor先生。

为什么使用这些航海术语？如果你还没猜到的话，我是一名爱好海事的史学家，2015年于美国军事大学取得军事史研究生（获得荣誉头衔）学位。尽管我的研究兴趣是1942-1943年间的南太平洋战争，但我同样对吸引其他人的主题感兴趣。

的确，我们发表的大部分内容与世界武装战争史相关，不过，我们的研究网络是广泛的，以期吸引尽可能多的读者。《军刀与卷轴杂志》第10卷第3期（2022年春季期刊）也是如此。

请在“奇怪却美妙”版块阅读我们的封面文章一由Aisha Manus撰写的获奖作品《Pexcho的一毛钱博物馆评论》。从干制首级（shrunken heads）到胀气药瓶再到维多利亚时期的医用（拷打）工具，该博物馆总有一款能为每个人带来娱乐或反感。甚至是Manus女士这一长期赞助商也对吞剑者瑞德的表演感到反感，以至于她需要请人来为他拍照。

作为历史缩影游戏协会—太平洋西南分会的合作伙伴，本刊很荣幸介绍2021年该分会的新晋史学家AJ Roth。评审员和本刊编辑对这位九年级自学者的印象极为深刻，以至于我们不得不把他的文章《战胜或战死的一天：1704年布伦海姆战役》收录在本期重要内容中。AJ，你太棒了！期待阅读你今年将撰写的作品。

本期收录的内容包括一系列优秀文章和书评：从古罗马哲学到自然选择、从第二次布匿战争到美国内战、以及1774年（美国革命最长的一年）合众国热气球驾驶员到好莱坞会见沙利文四兄弟。

最后，我想对欧洲目前的局势稍作评论，为本篇编者按划上句号。2022年2月28日，美国历史学会（AHA）发表声明“我们强烈支持乌克兰及其公民抵抗俄罗斯军事侵略，反对总统普京为其违反国际规范而捏造的扭曲观点”。尽管本刊将一直致力在内容和编辑立场上保持去政治化，但我们不能对AHA的呼吁置若罔闻“...普京把对历史的滥用作为袭击乌克兰的理由”（参见https://www.historians.org/news-and-advocacy/historians-condemn-russian-invasion-of-ukraine-(february-2022)）。

“战争是人类最悲惨、最愚笨的蠢行，无论是蓄意挑起战争，还是为其献计献策，这都是对全人类犯下的滔天罪行”，出自1947年6月3日德怀特·艾森豪威尔在西点军校毕业典礼上的演讲。

如果《军刀与卷轴杂志》员工被要求为AHA声明背书，我们将毫不犹豫地执行。

Ex Tenebris Lux

“光明来自黑暗”

Jeff Ballard

主编

加利福尼亚州亨廷顿海滩

2022年3月8日

The Historical Miniatures Gaming Society Pacific Southwest 2021 Essay Contest Rising Historian

This Day I Conquer or Die: The Battle of Bleinheim, 1704

Andrew Roth
Ninth Grade Homeschool

Abstract

Proportional to its effect on culture and history, the Battle of Blenheim is one of the most under-represented and under-researched events in military history. It saved the Grand Alliance between Holland, England, and Austria from being knocked out of the War of Spanish Succession and cemented the reputation of John Churchill, the First Duke of Marlborough (1650–1722), as a brilliant commander. In a few short hours, Marlborough had guaranteed the safety of Vienna—previously threatened by an army of 50,000 men—and finally brought the Grand Alliance to a true offensive footing. The glorious Sun King's court was taken over by emotional sobriety unseen in its 43 years since taking power: as historian James Falkner said, "When the dreadful news arrived, no one could credit the scale of defeat for French arms, and Louis XIV was so stunned by the news that it was thought at first that he had suffered a stroke." In sheer numbers, the Grand Alliance had captured an enormous amount of men and materiel that made sure that the French would not rebound for years. Camille d'Hostun de la Baume, duc de Tallard (1652–1728), the French commander, was imprisoned for seven years in England. This paper will provide support for the thesis that the Franco-Bavarians lost due to disunity of command, disunity of combat arms, and a lack of experience in their commanders.

Keywords: *En Muraille*, platoon firing, line, Reiter, canister/partridge, round shot, shell, artillery, Marlborough, Churchill

 doi: 10.18278/sshj.10.3.2

Este día venzo o muero: la batalla de Blenheim, 1704

Resumen

Proporcional a su efecto en la cultura y la historia, la Batalla de Blenheim es uno de los eventos menos representados y menos investigados en la historia militar. Salvó a la Gran Alianza entre Holanda, Inglaterra y Austria de ser eliminada de la Guerra de Sucesión española y consolidó la reputación de John M, el primer duque de Marlborough (1650–1722), como comandante brillante. En unas pocas horas, Marlborough había garantizado la seguridad de Viena, previamente amenazada por un ejército de 50.000 hombres, y finalmente llevó a la Gran Alianza a una posición verdaderamente ofensiva. La corte del glorioso Rey Sol se apoderó de una sobriedad emocional nunca vista en los 43 años desde que asumió el poder: como dijo el historiador James Falkner: "Cuando llegó la terrible noticia, nadie podía dar crédito a la magnitud de la derrota de las armas francesas, y Luis XIV estaba tan aturdido por la noticia de que en un principio se pensó que había sufrido un derrame cerebral". En números absolutos, la Gran Alianza había capturado una enorme cantidad de hombres y material que aseguró que los franceses no se recuperaran durante años. Camille d'Hostun de la Baume, duque de Tallard (1652-1728), el comandante francés, fue encarcelado durante siete años en Inglaterra. Este documento respaldará la tesis de que los franco-bávaros perdieron debido a la desunión del mando, la desunión de las armas de combate y la falta de experiencia de sus comandantes.

Palabras clave: *En Muraille,* tiro de pelotón, línea, Reiter, bote/perdiz, tiro redondo, proyectil, artillería, Marlborough, Churchill

战胜或战死的一天：1704年布伦海姆战役

摘要

与对文化和历史产生的影响成比例的是，布伦海姆战役是军事史上代表性被低估、研究最为缺乏的事件之一。这场战役挽救了荷兰、英国、奥地利大同盟，使其免于在西班牙王位继承战争中战败，并加强了第一代马尔博罗公爵约翰·丘吉尔（1650 – 1722）作为一名杰出指挥官的声望。短短几小时内，马尔博罗确保了维也纳的安全—此前维也纳受到五万名士兵的威胁—并最终将大同盟带到真正的进攻性地位。

路易十四的壮丽王宫笼罩在其43年掌权以来从未见过的阴霾下：正如历史学家James Falkner所说的那样“当可怕的消息传来，没人能相信法国军队的战败程度，并且路易十四因过于惊讶而一开始被以为中风了”。大同盟以绝对数量俘获了大量士兵和资源，确保法国在几年内无法崛起。法国指挥官Camille d’Hostun de la Baume, duc de Tallard（1652 - 1728）在英国被关押了7年。本文将提供证据证明，法国和巴伐利亚方面失败的原因包括指挥的不团结、作战军队的不团结、以及指挥官缺乏经验。

关键词：En Muraille，轮射（platoon firing），防线，-Reiter，榴霰弹（canister），圆弹丸（round shot），弹壳，马尔博罗，丘吉尔

The French Army

There is rarely an example in military history where an army backslid as much as the French army did in the leadup to the War of the Spanish Succession. Under the Marquis de Louvois (1641-1691), the French army had built up towering dominance over their enemies by the usage of innovative theory and the enforcement of discipline. Unlike most other European armies, the officer corps of the French army had taken a major interest in military theory. The question now became whether the army could back up its book smarts with victories on the field. This question was answered during the Thirty Years' War (1618–1648), as French armies repeatedly defeated their foes at battles such as Rocroi (1643), Freiburg (1644), Second Nordlingen (1645), and Zusmarshausen (1648). Further victories under the Bourbon banner would soon come during the Fronde, as Henri de la Tour d'Auvergne, Viscount of Turenne (1611-1675) battled Louis de Bourbon, Prince of Conde (1621–1686), and saved the royal family from capture more than once. King Louis XIV's (1638–1715) keen interest in the army had fostered its successes and royal support; he accompanied the army during the Fronde (1648–1653) and Dutch campaigns (1672–1678). During the Battle of Faubourg St. Antoine, Louis watched the battle from afar atop a hill, and campaigned in Brabant with Turenne and the Conde. He had also on many occasions spurred reform within the army, especially during the wars for the Low Countries.

This rapid improvement of the armed forces would at first appear to herald a future of French military dominance, but due to factors both internal and external, it did not. Some innovations, such as the militia system (which was essentially early conscription) were accidentally disadvantageous for the French.

The reason for this was religion. As the French army expanded, reaching 280,000 in 1678, they inevitably recruited Protestants, despite the French state being Catholic. Although some such as Turenne were extremely loyal, others were not quite as loyal. Frederick Schomberg, 1st Duke of Schomberg (1615–1690), and some 12,000 other troops would eventually flee to France's Protestant enemies, proving that religious divides in Europe ran deeper than mere doctrine. While alone not enough to destroy the structure of the army, there was no doubt that such a hemorrhage of men damaged it significantly and rattled the French to the core.

This major bloodletting of troops probably not only affected their manpower pool, but may have also caused "brain drain." Among the men who had fled the French army were 600 officers crucial to the development of military theory. French cavalry tactics were still remarkably primitive leading up to the War of the Spanish Succession, as they still tried to charge into battle as though they were Reiters, riding up to a target before halting and firing their weapons. While one cannot with complete certainty say that the cause of the slow progress after the French heyday was the desertion of so many officers who otherwise would have been valuable theorists and leaders, it is hard not to draw a correlation between the desertion and the brain drain.

During the first years of the War of the Spanish Succession, the French were hurrying their populace to the recruiting stations. The historian Rene Chartrand puts it nearly humorously: "Recruiting sergeants took almost anyone who could walk with a musket." They had even put militiamen into regular units, meaning that the French were now affected by religious, quality, and quantity problems. These were not mere growing pains: they were systemic problems that ran deep in the army, brought about by the higher command, which could only be mended by new reforms that never arrived.

Indeed, the weakness of the French army was uncovered in full by Prince Eugene of Savoy (1663–1736). Eugene was a military genius, among Napoleon's list of "great captains." During his campaign in Italy from 1701–1702, he repeatedly forced back several French armies nearly twice his size. Within scarcely a few months, the French were pushed back almost to Milan itself.

As stated previously, the issue of generalship was one of the major deciding factors of the battle of Blenheim. At the battle, the two French generals were Marsin and Tallard. Relative to their peers, Marsin and Tallard were completely new and thus unsuited to command of large forces, Marsin having never commanded more than 500 cavalry in his career. John Tincey has suggested the possibility that Marsin received his Marshal's baton mainly as a result of the praise of Tallard rather than his own skill. Despite his seeming competence, Marsin was inexperienced at a time when inexperienced generals' blunders could lead to catastrophe.

Marshal Tallard, commander of Franco-Bavarian forces at Blenheim. Unknown artist.

Tallard was the other French general at Blenheim. He was the primary commander of the French forces and was also relatively inexperienced. He had won his Marshal's baton at Speyerbach just a year before Blenheim, attacking a divided Allied force and bringing it to near annihilation. This was the primary catalyst for the seeming downfall of the Grand Alliance in southern Germany that year, directly leading to the Battle of Blenheim. However, he had his faults. The primary issue was his leadership style, something that would come to haunt him during the Battle of Blenheim. He was not particularly hard on his officers, as demonstrated at Blenheim, and preferred luxury to everything else. Thus, like Marsin, he was the wrong man for the specific job he was given.

Marlborough's Army

Even with the aforementioned defects in France's army, there was no reason to assume the French would lose at Blenheim. English forces under Marlborough had largely been halted the year before the Battle of Blenheim, being forced to give up plans to attack the Low Countries after losing thousands of Dutch troops in a botched attempt to take Antwerp. However, they had several major advantages.

Marlborough was friends with Turenne before his death, who was described as his "tutor in war" by Garnet Wolseley, 1st Viscount Wolseley (1833–1913). This means, ironically, that Marlborough's knowledge on how to defeat the French came from the French. He also had experience campaigning against the Moors, gaining an extraordinary amount of experience. It would have been rare to find a general quite as competent and experienced as Marlborough, who could draw on all sorts of experiences from the misty fields of England to the shores of Morocco.

But perhaps more importantly, he was not afraid to innovate. One of the most important innovations in 18th century military history was platoon firing, in which a battalion would be divided into four "firings," which would each fire on their initiative, effectively creating a monstrous wall of continuous musket fire. It would, like all other forms of volley fire, break down in the middle of battle, but in the early stages of a battle its value was priceless. Such a thing was not something that Marlborough would fail to use, and it would become a key part of English tactics, proving its worth consistently.

Another revolutionary but much less known aspect of English tactics was the *en muraille* formation. This would involve a cavalry force forming up into a wedge and packing each rank knee-to-knee before charging an enemy force. It was used at Ramillies, devastating the French cavalry and destroying their morale. As a rule of thumb, in the era of gunpowder, a densely packed wall of infantry or cavalry was enough to make enemy morale plummet. Another instance of a "wall" of troops advancing was at Mollwitz in 1741, where Kurt Christoph von Schwerin's (1684–1757) troops under Frederick the Great were described as a moving wall as they pushed ahead. The Austrians buckled under the weight of Schwerin's troops, withdrawing from the field in disgrace after what should've been an easy victory.

Lastly, Marlborough's companion in many of his battles, including at Blenheim, was Prince Eugene of Savoy. Although born in France, he was denied permission to join the French army, pushing him into the arms of the Austrians. Eugene was a brilliant commander, a "great captain" by Napoleon's standards. Not afraid of the smoke and din of battle, he had campaigned against the Ottomans in his early career, winning numerous victories for Austria, before fighting in the War of the Spanish Succession against the French in northern Italy. French armies, many two times bigger than his, repeatedly withdrew in his presence. In a few months, Eugene had maneuvered his way from Venice to the gates of Milan, delivering a heavy blow at Chiari to a French army twice his size.

Going into the campaign and battle, Marlborough had a massive advantage in leadership, and a large advantage in quality and tactics. In a vacuum, victory was certain, but if we look closer at the campaign, a different story unfolds.

Marlborough Begins His March: From the Netherlands to Donauworth

Marlborough had suffered a major defeat at the hands of the French as he attempted to attack Antwerp. However, Marlborough had also previously defeated the French, outmaneuvering them on several occasions, and defeating them in battle, taking 1,700 prisoners. Thus, his strategic defeat was by no means decisive.

What was truly troubling were events in the south. After the Battle of Speyerbach, the French seemed poised to assault along the Danube and strike into the heart of the Holy Roman Empire: Vienna. To many historians and contemporaries alike, it seemed like the Grand Alliance could fall with Vienna in 1704 or 1705. Unless Marlborough could relieve Vienna, all the blood and sweat over the past years of fighting in Europe would be in vain.

As Marlborough brainstormed a solution, relieving Vienna while si-

John Churchill, First Duke of Marlborough. Wikimedia Commons

multaneously keeping the Netherlands defensible was his primary goal. He wanted to draw the French armies after him, securing the Netherlands and allowing him to take his army from the Netherlands to the Danube,[1] where he would force the French into a battle. He would bring no siege train,[2] and thus would have to resort to unconventional means of forcing a decisive battle. This would culminate in his march to the Danube, in which Marlborough would purposefully expose his flanks to draw the French after him.[3] It was one of his most daring maneuvers, and one that would eventually culminate in the

Battle of Blenheim. Importantly, Marlborough also only had 14,384 British troops according to Robert Parker, although there were of course other men gathered along the way. [4]

In 1890, military historian and officer Theodore Ayrault Dodge wrote: "The one thing which distinguishes the great captains of history from the rank and file of commanders is that they have known when to disregard maxims, and that they have succeeded in disregarding them, and succeeded because of their disregard of them."[5] If we accept Dodge's opinion, then Marlborough was certainly one of the great captains. Perhaps no other general would be as bold as to purposefully expose their flanks, knowing full well it would make them a ripe target for an enemy army. In hindsight, Marlborough's move made complete sense. As he maneuvered like this, he could both protect the Netherlands by keeping French troops away and he could threaten French lines of communication in Austria, forcing them into a decisive battle on *his* terms. According to John Tincey:

> The letter makes Marlborough's strategy clear. His march to the east would draw French forces after him, leaving the Dutch with local superiority of numbers on their frontiers. The French in pursuit of Marlborough would head to defend the line of the river Moselle and by the time they realized that he was marching into Germany they would be far behind his army.[6]

Marlborough would carefully choreograph a dance around not just the French, but all military convention of the time.

Marlborough was no fool—that much is clear. On top of drawing French troops away from the Netherlands, forcing a decisive battle and spooking them away from Vienna, he would also consolidate his forces by bringing up Prince Eugene from the south and collect garrisons along the way, most notably at Coblenz.[7] Through his calculated tactics and strategies, Marlborough would bring together disconnected Anglo-Allied forces as the French tried to figure out where exactly he may have been going. Tallard asked Louis XIV what to do, and Louis simply said: "If the Duke marches, then so too must you march."[8] This was poor advice for two reasons: intelligence and clarity. Firstly, there was no clear path that Marlborough was taking. Maybe he was besieging Landau, maybe he was going to Ulm, maybe he was even trying to invade France. It was only discovered that he was taking the route to the Danube in the later stages of the campaign, just before the Battle of Schellenberg. Not even the enlisted men in the English army knew where they were going.[9] Secondly, saying essentially "just march" is not a tactically sound or clear plan. Tallard was left with no clear unified plan, and simply shadowed Marlborough's army while trying to unite with Marsin.

Thus, Marlborough had completely confused and disjointed the French command. He had not merely thrown a wrench in their plans; he had thrust a saber into their heart and thrown up smoke in their minds. The French pursuit was left even more confounded by an outbreak of glanders which killed and injured French horses, eventually forcing them to quarantine.[10] This was important as horses were expensive—by English measurements each cost 15 pounds, a small fortune in 1704. [11]

It seemed that the beginning of the campaign was going wrong for the French, and everything was going just right for the Allies. One anecdote shared by the Comte de Merode-Westerloo in his memoirs showcases this perfectly. On one night, the Comte was playing around with some other officers, trying to blow a foreign horn in the building where they slept. Eventually, after failing several times, they did blow the horn successfully; but this is where the trouble began. Farm animals heard the loud noise and naturally fled into nearby forests, but French troops also scattered into the forests, killing some of the farm animals in a showcase of bad discipline. The enraged locals skirmished with the French, managing to kill a large swathe of men, and causing great embarrassment in the French camp. It is possible that more casualties were inflicted during the crossing of the Black Forest due to this incident than during the actual pitched battles, if Merode-Westerloo is to be believed.[12] However, they did succeed in one thing, which is that they managed to get across the Black Forest without losing many troops, if any. After bluffing around the nearby Austrian forts, Tallard got his army through the Black Forest to unite with Marsin.[13]

This was important as it meant that the French armies were now united, and the Elector of Bavaria could wait for them to arrive as he avoided battle. Marlborough and Eugene were still divided, so this was crucial. But the Elector had different plans.

Fire and Sword: Marlborough in Bavaria

Nearby the fort-city of Donauworth—incidentally where one of the first incidents leading to the Thirty Years' War (the Donauworth Incident) occurred—the Comte d'Arco encamped his army, duly meting out assignments to his officers and setting up tents. Marlborough heard of this and moved ahead to attack, knowing it was an opportunity to strike before the French arrived. The Comte, despite being outnumbered, was ordered by the Elector to move to the Schellenberg. The Elector, so willing to evade Marlborough on every other occasion,[14] decided to give battle on this one day, whether out of a loss of nerve or simply a belief that he would win. Not wanting to waste a good opportunity, he encamped his forces in a position that made it seem as though he was resting for the night,[15] making the Franco-Bavarian commanders confident. However, Marlborough was aware that the more he waited the stronger the enemy position became[16] and he was not going to just let them entrench.

Around 6:00 PM, 50 men of the so-called "Forlorn Hope" were the first to assault the Bavarian position,[17] and

6,000 other men followed these brave troops. The first wave was intense, with M. de la Colonie writing:

> It would be impossible to describe in words strong enough the details of the carnage that took place during this first attack, which lasted a good hour or more. We were all fighting hand to hand, hurling them back as they clutched at the parapet; men were slaying, or tearing at the muzzles of guns and the bayonets which pierced their entrails; crushing under their feet their own wounded comrades, and even gouging out their opponent's eyes with their nails, when the grip was so close that neither could make use of their weapons.[18]

Was this exaggerated? Probably. However much it was exaggerated, there was an obvious grain of truth within it. M. de la Colonie was a veteran of over a decade at the time of the Battle of Schellenberg, so him stating something so resoundingly emotional was certainly saying something.

Eventually, the Allied soldiers stumbled back down the hill,[19] before renewing their assault later in the evening.[20] Importantly, on this second assault, Marlborough had knowledge of a line of whisker gabions that was almost completely unguarded on the Franco-Bavarian left flank.[21]

Marlborough, never one to give up a good advantage, began the second assault with this important intelligence in mind. While the French were pinned in place, the Allied soldiers began moving around the French flank, eventually encircling them. In the smoke and din of battle, the Bavarians did not realize the enemy infantry in their rear were Allied troops and refused to fire, believing they were reserves or reinforcements.[22] The French and Bavarians, now enveloped by fire and their opponents, initiated a futile last stand that lasted only long enough to be remembered in accounts of the battle. Casualties on both sides were immense: 9,000 Franco-Bavarian casualties, with 1,500 Allied deaths and 4,000 Allied wounded.[23, 24] This means that in total, of the 35,000 troops engaged on that day, a sum of 41% became casualties. In comparison, roughly 53% of all troops engaged at the Battle of Cannae became casualties. A quarter of the troops engaged at Shiloh became casualties.[25]

The consequences of the Battle of Schellenberg were dire for the Bavarians. Marlborough soon laid waste to the countryside—by his order[26]—forcing the Bavarians to disperse to protect their estates. In his own words, Marlborough wanted to "do our utmost to ruin his [the Elector's] country."[27] Propaganda probably exaggerated the amount of damage done, however. As M. de la Colonie writes, "Although I certainly found a few burnt houses, still the damage done was as nothing compared with the reports current throughout the country."[28] The effect of the raiding still was major, however, as the Bavarians dispersed in the hope of preventing Marlborough's men from doing further damage and importantly, not getting to the Elector's estates. Marlborough's goal was also to force the Franco-Bavarians into a battle, which would soon occur.

As Tallard advanced to unite with Marsin, Eugene closely shadowed him. He was unable, however, to make any major stands as his army was too weak to stand up to the French.[29] Events became dire as the French crossed the Danube at Lauingen and forced Eugene to retreat. Eugene wrote to Marlborough: "The enemy have marched. It is almost certain that the whole army is passing the Danube at Lauingen."[30] Marlborough read the letter and quickly rushed towards Eugene to aid him, realizing his chance for a decisive battle could come soon.

The joint Allied army encamped nearby the plain of Hochstadt. Dividing the plain were the Nebel Stream and the Danube River. These were major obstacles, as the Danube was impassable without a bridge, and the Nebel required a significant amount of construction labor to cross. But the Allies had one advantage, which was intelligence.

The Allies were accompanied by the "Old Dessauer," who was present at the same spot the battle was fought just a year before. He had fought another battle, deemed the "First Battle of Hochstadt," on the exact same ground. Thus, the Allies had some important level of intelligence on the terrain and how to cross, which was perhaps invaluable.

Another asset the Allies had was French overconfidence. As they had forced the Austrians under Eugene to retreat in front of them, and they had managed to force the Allies to work while the clock was ticking, they believed they were in a superior position. As Tincey writes, "When the Franco-Bavarians advanced to the north bank of the Danube they considered themselves to be facing an outmanoeuvred, demoralized and, to all intents, defeated Confederate army."[31] He is certainly right. On the skirmishes before the battle truly began, the Comte de Merode-Westerloo states:

> When I saw our troops falling back I also returned to the camp, and sat down to a good plate of soup in Blindheim along with my generals and colonels. I was never in better form, and after wining and dining well, we one and all dispersed to our respective quarters ... I don't believe I ever slept sounder than on that night. [32]

Then imagine the shock of the French officers when they realized the Allies were going to attack. Marlborough's cavalry were the first to rise to the attack: "the ... plain ... [in whole] appeared to be covered by enemy squadrons" as Westerloo put it.[33] Perhaps a large part of the reason was not by the fault of the French, but because the Allies had deceived them. From the skirmishes, several Allied soldiers had been captured, but these were under orders to tell the French that the Allies were going to retreat.[34] Thus, the French were unpleasantly surprised to discover that they had been lied to, and the Allies were not retreating, but advancing.

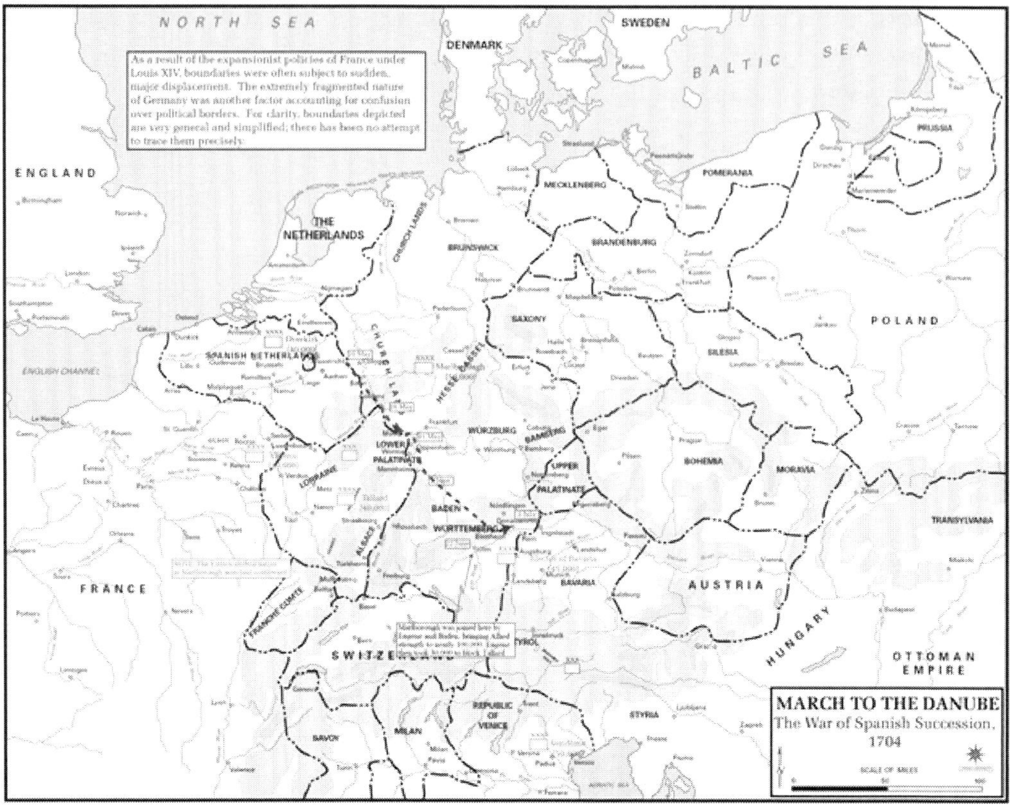

Marlborough's March to the Danube, 1704. Maps courtesy of the United States Military Academy Department of History.

The First Blow is Struck: Marlborough Attacks

Caught by surprise, the French position was now perilous. They did have a series of breastworks nearby the village of Blenheim, yet they were divided in the sense of tactical opinion. As Marsin, the Elector, and Tallard observed the situation, they decided to head to the Church tower nearby Blenheim and call together a council of war. Tallard believed that the French cavalry were best suited to stay behind the Nebel stream and should have charged down towards the stream once the Allies crossed, while Marsin and the Elector both said they should contest the crossing directly.[35]

Rather than both sides uniting, they all went their own way. As Napoleon once said, "if you weaken your means by dividing your forces, or break the unity of military thought … you will have lost the most favorable occasion."[36] No greater example of this can be found than the Battle of Blenheim.

Either way, Marlborough was now advancing. Tallard and the Franco-Bavarians made new fortifications by throwing up unmanned wagons, cutting down trees and vineyards to give better views, etc. On Tallard's side, they had failed to secure the narrow area be-

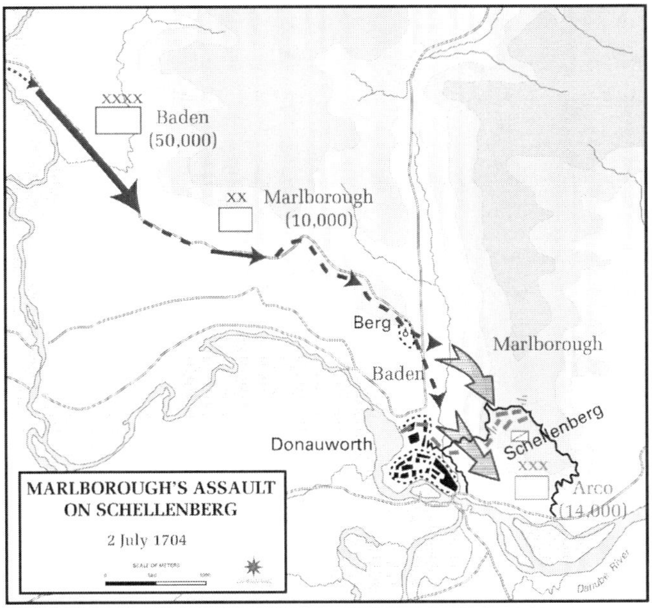

The Battle of Schellenberg, 2 July 1704. Maps courtesy of the United States Military Academy Department of History.

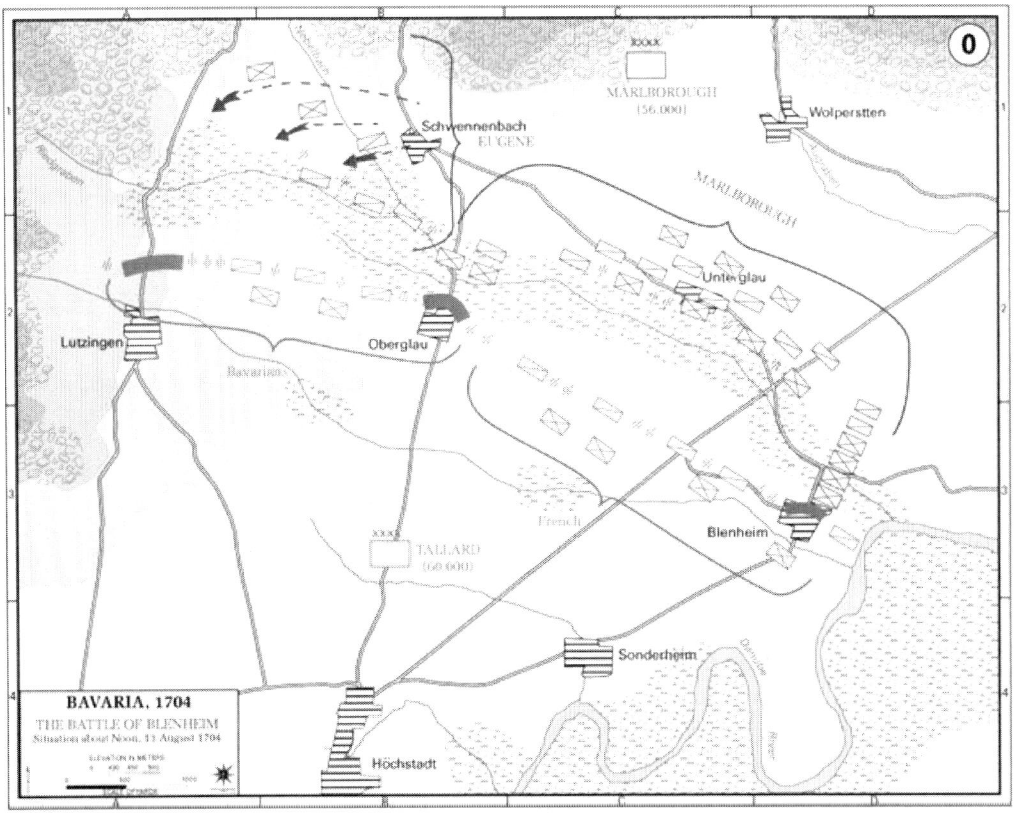

Battle of Blenheim. The two armies in position, noon of August 13, 1704. Maps courtesy of the United States Military Academy Department of History.

tween the marshy Nebel and the Fuchsberg hill. This was crucial ground, but Marlborough was opposed by practically no French troops.[37] Next on Marlborough's list was to attack Blenheim itself. He gathered near Unterglau, although the area was soon lit ablaze by French gunners.

French forces at Blenheim were placed like so: nine battalions manned the village itself, while three stayed in the nearby cornfields, and eleven were placed behind them.[38] Marlborough was at a disadvantage, as he was attacking an army in position, but he still could rely on tactical superiority and surprise to defeat the Franco-Bavarian army.

Although Marlborough himself was in position, his right wing under Prince Eugene was not yet ready. It would take several hours for him to be organized, and 30 minutes after Eugene told Marlborough he was in line at 1:00 PM, Marlborough gave the order to attack.[39]

John Cutts's men would be the first to attack. One of his brigadiers, Rowe, gave the order that no man would or should fire until he had struck the first blow with his sword on the outer French breastworks.[40] The French showed proper fire discipline, as they waited and waited until the opportune moment to fire. Rowe, having given an order that could not be followed through, was repulsed on his first attack. In fact, he was taken in the flank by the elite *Gens d'Armes* of the French cavalry, who nearly destroyed his brigade completely. At one point, the Allies lost a colour[41] that had to be recaptured later in the battle. However, the Allies had a trick up their sleeve. Nearby, on the French flank, Allied troops lied in wait. These sprung upon the *Gens d'Armes*, who routed, seeing they were outflanked and nearly completely encircled.

Of course, the French troops nearby were demoralized by this occurrence, but even more important was the demoralization of the French command. Tallard tried desperately to rally the *Gens d'Armes*, failing consistently, watching as his greatest squadrons melted away from the field.[42]

Marlborough met with Cutts and ordered him to stop his attacks. But the brainless commander decided to attack once more, driving the French from the outskirts of Blenheim before being repulsed for the second time.[43]

While the Allied strength was never enough to attack Blenheim successfully, it was enough that the French commander who was stationed in Blenheim decided to stack up 12,000 troops on the area, who never moved other than to counter-attack. Thus, 5,000 Allied men pinned 12,000 French and Bavarian men.

The Bloody Fields of Hoechstaedt: Oberglau

With Cutts launching his final attack for just the moment, the fighting around Blenheim had for the moment ceased. Although the French made a few uncoordinated counterattacks, these never got far. What was becoming the primary focus of the battle was the center. There was barely any infantry support for the French cavalry in the center, and Marlborough was just the general to exploit this. He now put into effect his primary tactic: shifting the center of focus of the battle. Previously, the battle had been centered on Blenheim, but now Marlborough wanted to shift it to Oberglau in a maneuver that would truly show his genius.

He began his attack. The Danish cavalry surged forth, but try as they might, they could never truly get lodged into the Franco-Bavarian positions. Shortly afterwards the Prince of Holstein-Beck entered the fray, advancing with two brigades of Dutch infantry but being repulsed, himself becoming mortally wounded. Allied forces were consistently harassed in their flanks, and it seemed the collapse of their center was imminent.

Marlborough did not panic, however. He moved up nine cannons loaded with canister shot and placed some of his elite cavalry in a position to threaten the flank of any attacking French forces. Not wishing to share the fate of the *Gens d'Armes*, the French cavalry never attacked. Marlborough's center was secure.

The Comte de Merode-Westerloo charged with his cavalry across the Nebel stream, breaking through several Allied lines. However, he soon was pushed back and disorganized by a "third line," his troops now tired. This left the Allies in a position to charge the French positions, as they were now tired, while the Allies had not even committed all their reserves yet. But before that attack, I would like to go into detail on the events on Eugene's flank.

Lutzingen: Eugene's Side

With Marlborough holding the center and pinning the left, one may wonder how Eugene was doing. Ever since the battle began, he had been launching a series of attacks, which had also pinned the French in his sector. Eugene had the Old Dessauer in his ranks, who was not afraid to launch several attacks on Lutzingen, the primary town/anchor point of the French flank. He brought up four Prusso-Danish brigades, who began their assault quickly. French canister and musketry dealt horrendous damage to the Dessauer's men, while his cavalry tried desperately to hold against French charges. Eventually, he was repulsed, losing ten colours as his flank caved in.[44]

The Old Dessauer's cavalry was essentially removed from any further attacks, and so once he attacked again, he was repulsed again due to having barely any cavalry support. Despite this— being disorganized and tired—the French were unable to follow up their success, meaning that the fighting on this side was also essentially stagnant. Crucially, Eugene had pinned the French troopers in place, meaning that when Tallard requested cavalry support from Marsin he was rejected. Marlborough was now set for the decisive attack.

The Winning Stroke: Oberglau Again

Marlborough was in the perfect position. The French cavalry was tired and unsupported, while both French wings were now pinned. He still had not committed his full reserve, made up of cavalry and infantry, and could now begin the winning stroke. At 5:00 PM, 8,000 Allied cavalry clopped forth. Now the moment that the winners or losers were decided had come.

The infantry and cavalry of the Allied army were never unsupported, making them infinitely superior in a tactical sense to the unsupported and tired French cavalry. This was incredibly valuable at Blenheim, as the sort of combined arms of the Allied army had a field day running over the French cavalry. They were quickly defeated, while Blenheim and Oberglau were encircled. Oberglau formed a bottleneck which the French and Bavarians still controlled, and thus to cross the stream Marlborough's only option was to encircle and blockade it. Tallard's flight had also exposed the flank of Marsin, who had to withdraw with the Elector from the field after a couple more hours of fighting.

Tallard was taken prisoner after a few Hessian soldiers recognized him while he was trying to escape. A conversation ensued between Tallard and Marlborough once Tallard got to Marlborough's coach, as Tallard asked Marlborough if he could order his troops to withdraw from Blenheim. Marlborough refused, replying that he was in no position to ask.

Another part of the conversation is recorded by Tincey. "When the firing was over, the Duke asked Tallard, how he liked the army? He answered, with a shrug, 'Very well, but they have had the honour of beating the best troops in the world.' The Duke replied readily, 'What will the world think of the troops that beat them?'"[45]

The Battle Ends: The (Short) Siege of Blenheim

Marlborough turned his sights to Blenheim. As there were no threats from either the center or the right flank of his army, he could entirely focus on that one area and immediately sent more artillery and troops there. Several batteries were placed outside of Blenheim, which opened fire as the town soon caught ablaze. As French troops tried to escape, they were consistently gunned down, with no escape route to speak of.[46] When darkness fell, 10,000 French soldiers fell with it.

Aftermath

There were 40,000 French casualties in total. Of these, 14,000 were prisoners, and 20,000 were wounded or killed. Amongst the fallen was Tallard's son, who was killed in action, along with many other French officers and soldiers. Emotionally shattered in defeat, 6,000 hopeless Franco-Bavarian refugees deserted. Some 3,600 tents,

7,000 horses, 5,400 wagons, 100 guns/mortars, 129 colours, and 110 cavalry standards were captured.[47] However, it is not as if the Allies did not suffer great casualties as well. Eugene lost 5,000 troops, Marlborough 9,000.[48] Even so, the losses were trivial compared to the reward.

Marlborough had, as mentioned earlier, not taken a siege train with him. But this was of trivial importance after the battle ended. After capturing the French artillery, and obliterating their army, French garrisons that had been left behind after the conquest of several forts were not particularly eager to resist. He soon captured several forts, undoing almost all the work the French had done the year before.[49] It was the textbook definition of a decisive victory, as French forces before him continually capitulated.

Marlborough's victory was celebrated across England. Near Westminister hall, French standards were paraded around to demonstrate the victory, and Tallard was sent back to Nottingham in captivity. "How could God do this to me after all I have done for him?" Louis XIV asked, after hearing the news of the defeat.[50]

Conclusion

There are several things to draw from the Battle of Blenheim. The two most important are, in my opinion:

1. Unity of command
2. Combined arms

While there is a series of other things to think about, these are the primary controllable issues. As mentioned earlier, Napoleon once said "if you weaken your means by dividing your forces, or break the unity of military thought ... you will have lost the most favorable occasion." Immediately, once Marlborough began his attack, Tallard's army and Marsin's army split. As Tincey wrote: "The army of Tallard was drawn up quite separately from the Franco-Bavarian forces commanded by Marsin and the Elector."[51] This is certainly true, and a large part of the reasoning behind this is that Tallard and Marsin may have had a rivalry, which was a mistake on any day, but catastrophic when facing a unified army such as that of Marlborough's.

With a divided army, neither wing could draw upon the support of the other. At one point in the battle, Marsin refused to transfer his cavalry to Tallard, despite Tallard's requests. As Falkner says, "Tallard ... recommended that Marsin detach some of his army to reinforce the right wing. The younger Marshal, who despite the fears of infection had already sent part of his cavalry to help Tallard earlier in the day, now refused."[52] Although it is unlikely that disunity of command was the only reason behind the refusal, it was certainly a large part of it. Thus, by dividing their army along personal lines in such a dire moment, the French and Bavarians had doomed themselves unnecessarily.

Perhaps equally or more importantly the French did not have proper support from every combat arm at every moment. While near Lutzingen, it was the Allies who did not have proper cavalry support, in every other sector it was the French and Bavarians, who would either only have cavalry or only have infantry. This was caused by a supposed "mental breakdown" (as Falkner puts it) of one of the French commanders, who sent most of his infantry troops into Blenheim, isolating the French cavalry on the plain of Hochstadt. This meant that when the cavalry charged and charged, they were the only combat arm involved in any engagement. The dirty work went consistently to the cavalry, who were by the end of things tired, demoralized, and disorganized. In comparison, every Allied unit near Oberglau or Blenheim had the ability to draw on cavalry and artillery support, and by the end of the battle, many had not even been engaged.

A lost opportunity occurred in the center when the French and Bavarian cavalry failed to charge the Allied forces who were already beaten. If, perhaps, the French and Bavarian cavalry found themselves supported by artillery and infantry, they would have been able to make it across the Nebel and would have split the Allied army in two. This side effect of lack of support may have been one of the biggest reasons the French lost.

Not even Jomini would be able to predict the results. Marlborough's campaign ran contrary to his belief in always keeping lines of communication/operation secure, and it seems as though Jomini may not have had a firm grasp on the battle. He claimed that Tallard shouldn't have attacked along the Danube and shouldn't have stayed entrenched, when it was clearly Marlborough who attacked along the river.

Nonetheless, Marlborough did exercise some of the principles of war that Jomini laid out, such as attacking the decisive point in a concentrated way. Once Marlborough had successfully pinned Tallard's flanks, he moved towards the decisive point (that being the Nebelbach) and successfully crossed it, utilizing his reserves and his available troops to that end.

The way in which Marlborough carried out the battle was worthy of Napoleon. The flexibility of Marlborough's commands might as well have been out of the Battle of Austerlitz or the Battle of Jena. I would go so far as to argue that, in this specific context of flexibility, Marlborough and Napoleon were both equal. Napoleon saved Ney, Marlborough saved the Prince of Holstein-Beck.

An interesting statistic is the difference in artillery between Marlborough and Tallard. While the numbers are different across all sources, there is a consensus that the French had a massive advantage over Marlborough and Eugene. Still, Marlborough managed to apply his artillery at the right moment and the right position, such as during the French counter-attack in the center. This limited but decisive usage of artillery would've been remarkably impressive for any general, not just Marlborough. Thus, despite not having a superiority in firepower through artillery, he used it at the right time in the right area, making up for the difference.

Marlborough also had a major effect on his troops and their morale. At the opening of the battle, Robert Parker (present at the battle) noted Marlborough accepted several religious symbols handed to him and exclaimed: "This day I conquer or die." For it to be mentioned in a triumphalist way in a memoir of one of the troops present, it is almost certain that it had a positive effect on the morale of the troops. After the battle, too, Marlborough was remarkably more trusted than before. At the Siege of Bouchain, before Marlborough's assault, Parker also stated:

> I must confess I did not like the aspect of the thing…[but] he would not push the thing unless he saw a strong probability of success; nor was this my notion alone; it was the sense of the whole army, both officer and soldier, British and foreigner. And indeed we had all the reason in the world for it; for *he never led us on to any one action, that we did not succeed in.*[53]

Therefore, the French and Bavarian loss was due to a laundry list of factors, but chiefly their disunity of command and their lack of supporting combat arms in most sectors. It's not as if the French and Bavarians didn't put up a fight: of the 108,000 troops that were engaged, 44% became casualties, proportionally more than either the battles of Schellenberg or Shiloh. The statistics and the decisive moments of the battle perhaps show why the French thought Marlborough won by luck and not by skill. A few mistakes mended here and there may have been the difference between a decisive victory and a catastrophic defeat. Those mistakes committed turned into blunders that eventually turned into disaster. Many decisive battles end with a bang, but this one did not. It ended with a nuclear explosion large enough to shatter the myth of French invincibility under their glorious Sun King. Gone were the days of French martial superiority scattered like English musket-balls on the field of Hoechstaedt. France could no longer lay claim to the glorious legacy of Conde or military genius of Louvois. Turenne was killed by a cannon ball in 1675, and so were any feelings of French martial superiority over Europe on those August days in 1704. Instead, the enduring legacy of the legendary Duke of Marlborough was forged in blood and iron.

Works Cited

Chartrand, Rene. *Louis XIV's Army.* Osprey Publishing, 1988.

Colonie, M. de la. *The Chronicles of an Old Campaigner.* Leonaur, 2012.

Dodge, Theodore Ayrault. *Alexander: A History of the Origin and Growth of the Art of War.* Boston: Houghton Mifflin Harcourt, 1890.

Falkner, James. *Marlborough's War Machine.* Pen and Sword, 2014.

———. *War of the Spanish Succession 1701-1714.* Pen and Sword, 2015.

Longueville, Thomas. *Marshal Turenne.* Alpha Editions, 2019.

Lunde, Henrik O. *A Warrior Dynasty: The Rise and Decline of Sweden as a Military Superpower.* Casemate, 2014.

Luvaas, Jay. *Napoleon on the Art of War.* Free Press, 1993.

Miller, Donald. *Vicksburg: Grant's Campaign that Broke the Confederacy.* Simon and Schuster , 2019.

Nolan, Cathal J. *Wars of the Age of Louis XIV, 1650-1715.* Westport: Greenwood Press, 2008.

Parker, Robert, Merode-Westerloo, and David G. Chandler. *Robert Parker and Comte de Mérode-Westerloo: The Marlborough wars (Military memoirs).* Archon Books, 1968.

Notes

1 John Tincey, *Blenheim 1704: The Duke of Marlborough's Masterpiece,* 30.

2 Ibid., 36.

3 James Falkner, *Blenheim 1704: Marlborough's Greatest Victory,* 44.

4 Robert Parker, *Memoirs of Robert Parker,* 30.

5 Theodore Dodge, *Alexander: A History of the Origin and Growth of the Art of War,* 4.

6 John Tincey, *Blenheim 1704: The Duke of Marlborough's Masterpiece,* 30.

7 James Falkner, *Blenheim 1704: Marlborough's Greatest Victory,* 43.

8 Ibid., 40.

9 Ibid., 43.

10 John Tincey, *Blenheim 1704: The Duke of Marlborough's Masterpiece,* 32

11 James Falkner, *Marlborough's War Machine,* 173

12 Merode-Westerloo, Parker and Chandler, *Robert Parker and Comte de Merode-Westerloo: The Marlborough wars (Military memoirs),* 160

13 John Tincey, *Blenheim 1704: The Duke of Marlborough's Masterpiece*, 28.

14 James Falkner, *Blenheim 1704: Marlborough's Greatest Victory*, 49.

15 John Tincey, *Blenheim 1704: The Duke of Marlborough's Masterpiece*, 32.

16 James Falkner, *Blenheim 1704: Marlborough's Greatest Victory*, 53.

17 John Tincey, *Blenheim 1704: The Duke of Marlborough's Masterpiece*, 33.

18 M. de la Colonie, *Chronicles of an Old Campaigner*, 185.

19 James Falkner, *Blenheim 1704: Marlborough's Greatest Victory*, 61.

20 John Tincey, *Blenheim 1704: The Duke of Marlborough's Masterpiece*, 33.

21 James Falkner, *Blenheim 1704: Marlborough's Greatest Victory*, 62.

22 Ibid., 63.

23 Ibid., 67.

24 John Tincey, *Blenheim 1704: The Duke of Marlborough's Masterpiece*, 36.

25 Donald Miller, *Vicksburg: Grant's Campaign that Broke the Confederacy*, 84.

26 James Falkner, *Blenheim 1704: Marlborough's Greatest Victory*, 72.

27 William Coxe, *Memoirs of the Duke of Marlborough*, 210.

28 M. de la Colonie, *The Chronicles of an Old Campaigner*, 207.

29 James Falkner, *Blenheim 1704: Marlborough's Greatest Victory*, 81.

30 John Tincey, *Blenheim 1704: The Duke of Marlborough's Masterpiece*, 37.

31 Ibid., 39.

32 James Falkner, *Blenheim 1704: Marlborough's Greatest Victory*, 94.

33 Ibid., 100.

34 Ibid., 95.

35 John Tincey, *Blenheim 1704: The Duke of Marlborough's Masterpiece*, 48-49.

36 Jay Luvaas, *Napoleon on the Art of War*, 89.

37 James Falkner, *Blenheim 1704: Marlborough's Greatest Victory*, 92-93.

38 Ibid., 106.

39 Ibid., 120.

40 John Tincey, *Blenheim 1704: The Duke of Marlborough's Masterpiece*, 53.

41 Ibid., 56.

42 James Falkner, *Blenheim 1704: Marlborough's Greatest Victory,* 139-140.

43 Ibid., 124.

44 James Falkner, *Blenheim 1704: Marlborough's Greatest Victory,* 127.

45 John Tincey, *Blenheim 1704: The Duke of Marlborough's Masterpiece,* 90.

46 James Falkner, *Blenheim 1704: Marlborough's Greatest Victory,* 163-166.

47 Ibid., 178.

48 Ibid., 177.

49 John Tincey, *Blenheim 1704: The Duke of Marlborough's Greatest Victory,* 88.

50 Cathal Nolan, *Wars of the Age of Louis XIV, 1650-1715.*

51 John Tincey, *Blenheim 1704: The Duke of Marlborough's Masterpiece,* 49.

52 James Falkner, *Blenheim 1704: Marlborough's Greatest Victory,* 139.

53 Robert Parker, *Robert Parker and Comte de Merode-Westerloo: The Marlborough wars (Military memoirs),* 108. [Emphasis added.]

No Faith in Hope: Darwin, Lady Hope, and the Evolution of an American Lie

Michael Kahn

Retired Police Officer, BA in American History, SUNY-Empire State College

Abstract

Charles Darwin did not recant his work on evolutionary biology on his deathbed. Rather, it is an apocryphal story, with some basis in fact, which became distorted with retelling over time, as well as utilized by polemicists and proselytizers to discredit Darwin.

Keywords: Charles Darwin, Lady Hope, evolution, deathbed, evangelicals, affidavit, converted, myth, biology, interview

Sin fe en la esperanza: Darwin, Lady Hope y la evolución de una mentira estadounidense

Resumen

Charles Darwin no se retractó de su trabajo sobre biología evolutiva en su lecho de muerte. Más bien, es una historia apócrifa, con alguna base de hecho, que se distorsionó con el tiempo y fue utilizada por polemistas y proselitistas para desacreditar a Darwin.

Palabras clave: Charles Darwin, Lady Hope, evolución, lecho de muerte, evangélicos, declaración jurada, convertido, mito, biología, entrevista

不相信霍普：达尔文、霍普夫人以及一个美国谎言的演变

摘要

查尔斯·达尔文在临终前并未公开宣布放弃其进化生物学著作。相反，这是一个虚构的谣传，其中存在部分事实，但随着不断复述而被曲解，同时被辩论家和宗教劝导者用于质疑达尔文。

关键词：查尔斯·达尔文，霍普夫人，进化，临终床，基督教福音派教徒，宣誓书，信仰改变，谣传，生物学，访谈

doi: 10.18278/sshj.10.3.3

Contrary to popular belief, Charles Darwin did not "invent" evolution or the theory of evolution. Darwin did not claim that human beings evolved from monkeys. These and other claims, also better understood as miscomprehensions of the central elements of evolutionary biology, have been conventionally promulgated by religious fundamentalists attempting to undermine the validity of scientific facts. In an effort to enforce the power of Abrahamic creation theology/mythology, attacks on Darwin (and his works) have been the focal point of their evangelistic strategy. One lasting claim is that Darwin recanted his scientific work and converted to Christianity on his deathbed. Apparently, none of those promulgaters repeating this tale can even cite their respective sources, nor can those who claim this story has no factual merit name the origin of the source of this tale.

For the purposes of this critique, a major fundamentalist claim about the elderly and dying Darwin will be examined based on what probably occurred. There is no reliable evidence and or testimony to substantiate Charles Darwin renounced the validity of his work on his deathbed and/or converted to Christianity. So where does this story come from and why? About a century ago American evangelicals exploited the false story that Charles Darwin recanted his scientific work to Lady Elizabeth Hope in order to discredit the teaching of evolution. Born Elizabeth Reid Cotton in Longford, Tasmania, Lady Hope (1842–1922) married into the nobility in her first union with Admiral Sir James Hope, and after his death, to a Thomas Anthony Denny. After years of dedication to work in temperance and evangelism, she came to America in 1913 and began spreading her story of her alleged encounter with Charles Darwin.[1] It is unknown what the original motivation was for her to spread this story, but at the least, she could use her fame associated with her account of Darwin as a platform to continue her work. Later on, others would exploit her story for their own nefarious purposes.

Although not the first account about Darwin, Lady Hope's account was the one that has left a popular belief amongst the evangelical culture to this day. Nor were stories of deathbed conversions or revelations original to Darwin, as they seem to have generated around anyone of celebrity-like status in which said stories would serve the proselytizing purposes of the evangelicals.[2] For instance, there are deathbed conversion stories about Charles II, Washington, Oscar Wilde, and even John Wayne.

It is not a coincidence that Lady Hope's personal financial situation was pretty dismal at the time her account of her interview with Darwin gains mass attention. She had a reputation for being a spendthrift, and she had gained notoriety in 1911 and 1912 for her bankruptcy proceedings, apparently blowing through all the money left to her from both her deceased husbands. Hope also had become estranged from her stepsons, in part because she still went publicly by the noble name and title from her first husband. Hope then relocated to the East Coast United States, in the hopes of starting over.[3] 1915 would be the pinnacle year for her, as this is where her place in history truly intersected with Darwin's, and the legacy passed down to evangelicals today.

Darwin and Lady Hope may have met. Furthermore, Darwin probably praised her efforts on temperance, as she was the leading crusader of the movement, and the Darwins, supporters of her social work.[4] However, supporting the "quality of life" work done by missionaries is not the same as supporting their theological doctrines. A biased and subjective mind can easily conflate the two overlapping elements of acts and beliefs.

Lady Hope and her story first came to public attention on August 4, 1915, at the Northfield Seminary girl's prep school in East Northfield, Massachusetts. Lady Hope was a guest speaker along with Professor A.T. Robertson (of Southern Baptist Theological Seminary in Louisville, Kentucky). On the previous day, Robertson had given an exegesis on the New Testament Epistle to the Hebrews which he developed into an admonition on Darwinism. In private the next day, Lady Hope confided in Robertson, and others there, including an editor from the Baptist *Watchman-Examiner*, her account of how she personally knew Darwin and Darwin's personal affection for the Epistle to the Hebrews.

> He [Darwin in 1881] was almost bedridden for some months before he died … I made some allusion to the strong opinions expressed by many persons on the history of the Creation, its grandeur, and then their treatment of the earlier chapters of the Book of Genesis. [Darwin] seemed greatly distressed, his fingers twitched nervously, and a look of agony came over his face as he said: "I was a young man with uninformed ideas. I threw out queries, suggestions, wondering all the time over everything; and to my astonishment, the ideas took like wildfire. People made a religion of them … I have a summerhouse in the garden, which holds about thirty people … I want you very much to speak there …"
>
> "What shall I speak about?" I asked.
>
> "CHRIST JESUS!"[5]

There are several elements here that, at face value, may have a hint of fact in them. Simultaneously, the claims made by subsequent promulgators of this tale do not hold up against the original transcription. First and foremost, Darwin is not dying or on his "deathbed," as Darwin died in 1882 and this interview allegedly took place the previous year. Secondly, Darwin does not make any statements that he was converting or recently had converted. Thirdly, there is no clear statement that he renounced anything, specifically his theory of natural selection. Yet a closer look at the wording reveals euphemisms for conversion and renouncement.

"Distressed" appears to be another way of saying Darwin regretted proposing his theory and therefore undermining the literal account of creation in Genesis. Then the reference to Darwin's confession to being "young" and "uninformed" are also intended to have Darwin demoralize his own confidence in his work. The last part of the story, where Darwin solicits Lady Hope to preach about "Christ Jesus" is substituted for him actually verbally accepting the literal authority of scripture, as well as accepting the evangelical orthodoxy. Darwin does not outright convert or renounce, but he does allegedly say this in a more passive-aggressive language.

This, however, is just the first version of the story Lady Hope apparently gave that caught the attention it did. The story was repeated in multiple Christian publications up until the late 1920s, as well as the subject of the "Los Angeles Affidavit," which was also the product of religious leaders to whom Lady Hope told the story. The affidavit was not just an anomaly that the fundamentalists wanted to document—they were hoping to utilize it in their crusade against the public teaching of evolution.[6] Their social warrior was none other than three-time failed Presidential candidate William Jennings Bryan (1860–1925). Bryan was already on the offensive against public educators.[7] In order to gain Bryan's attention, the story was transcribed into an affidavit in order to vouch for its factual accuracy by Annette Parkinson Smith, dated June 7, 1922. Smith's letter to Bryan (excerpt below) has some interesting deviations and embellishments from the Hope account as given in the *Watchman-Examiner*:

> "Oh"! [Darwin] replied "those theories of evolution! Oh, I put out those theories when I was a young man, searching, searching for knowledge, and they made a religion of them! Oh! if I could only undo them"! and Dr. Darwin appeared much agitated as he said these words, evidently thinking that his hypothesis had done great harm. He expressed himself similarly on another occasion so that Lady Hope was quite convinced he repudiated them … but Mrs. [Emma] Darwin, and her sons, knowing that the fame of the family rested upon said evolutionary theories,-and, probably, as one remarked to me recently,-the income from the sale of those books-did not care to welcome a visitor who brought a gospel message, and Mrs. D[arwin] showed some little (polite) displeasure when present in the room, and the gospel meeting was never held in the summer house.[8]

One of the cosigners of the affidavit accompanying Smith's letter was Professor Melville Dozier (1846–1936), formerly an administrator of the college which would become UCLA and a former assistant superintendent of Los Angeles public schools. Dozier's name as the chief signatory was meant to give some intellectual credence to the claim. Smith even cites Dozier's credentials in the letter as a "scholar and a scientist."[9]

Charles Darwin (1809-1882), circa 1854, Wikimedia Commons. Darwin was born on the same day in history as Abraham Lincoln. Darwin's contributions to the sciences, namely evolutionary biology, would become synonymous with his name.

Emma Darwin (1808-1896), portrait painting by George Richmond, Wikimedia Commons. Emma was actually a first cousin to Charles. The Darwins would have ten children together. Emma was present at Darwin's deathbed, unlike Lady Hope.

Furthermore, she closes the letter with "Every effort is needed to counteract teaching of unBiblical theories in L.A. schools, High Schools, University & c."[10] There are several problems with the story attested to in the affidavit. The story now has Lady Hope meeting with Darwin several times, not just once. This time the story has Darwin clearly mentioning evolution by the term "evolution." Additionally, Emma Darwin was not included as being present in the original account as given to the *Watchman-Examiner*. These are all significant elements that cannot be reconciled simply as Hope remembered more pertinent details as time went along. Whether the details got lost or "improved" with transmission cannot be known with certainty. Yet the closing remarks are telling in that it was obvious Smith and her group sought to exploit this story and pass it along to a man of Bryan's notoriety and celebrity influence. In addition, it was a desperate inclusion of Prof. Dozier as a cosigner in the letter, seeing as how he was not attesting to being present when Darwin supposedly had this conversation with Lady Hope.

In response to Smith's letter, Bryan stated he was familiar with Lady Hope's account, but that it (as articulated in the letter to him) was insufficient to refute the evidence documented in his works. Furthermore, he wrote "The question is not whether Darwin at the last moment came back to Christianity—as Christians we believe we can believe this possible—*but unless he specifically repudiated his views* as to the origin of man..." (my emphasis).[11] At first glance, this looks as though Bryan outright rejected the story as beneficial to his cause. However, the wording in his reply seems to be a subtle implication that those attesting to in the affidavit should improve their memory of Lady Hope's account, and then revise their statements so that the account sounded more convincing. The reason this seems likely on Bryan's mind and explained in his response is that Smith has Lady Hope claiming Darwin wished he could "undo" his theories of evolution. Furthermore, Smith goes on to claim that Lady Hope was "quite convinced he [Darwin] repudiated them [evolution]."[12] In physical terms, Bryan saw the potential of the story, but currently, there was nothing kinetic and, therefore, not utilitarian. Bryan was an intellectual and an attorney-at-law and he would have known that this hearsay, especially so long after Darwin's death, was of no value in a court of law. Then again, Bryan was also fighting in the court of public opinion, where the rules of evidence are quite different.

Next, we come to an actual professor of college biology at John Fletcher College in Iowa, Dr. S.J. Bole. In 1940, Bole published *The Battlefield of Faith* and in it, the account Lady Hope gave him of her meeting with Darwin. Bole mentioned that his undergraduate professors all accepted evolution as factual allegedly ignoring evidence against it, which he does not cite or list. Bole goes on to comment that when he was teaching at Wheaton College, he was emotionally moved by an evangelist who shared his "born again" experience after his religious convictions were destroyed by his higher learning.[13]

This version of the Hope story is the longest and most nuanced of all previous versions. According to Lady Hope, she had to flee England of religious persecution at the hands of the Darwin family and their partisans. Hope claimed that she did not remember what years she met with Darwin, but this time, Darwin strongly regrets his theories conflicting with the Creation story in Genesis. Bole concludes the chapter on Lady Hope claiming there are other scientific men who believe in evolution, but that: "There were and are others equally scholarly that did not and do not believe in evolution ... I often ask my students at the end of a semester' study ... to carefully give the evidence in favor of evolution. With few or no exceptions, they find it difficult ... every such course shows that nature and life on every hand speak of a Creator and a world of order."[14] The sum and substance of this version are very similar to the version as printed in the *Watchman-Examiner*. The emphasis this time seems to rely on Darwin's regret (or recantation). Bole claims there are other scholars who did not accept evolution but never mentions a single one. The claim by Hope that she was being persecuted is suspect: were these her words or Bole's? This sounds more like Bole is reflecting on his position now amongst men of science, or, in other words, the removal of the divine from the life sciences and classrooms is persecution against fundamentalist Christianity. And what better story to sell than one of a noble English lady being persecuted for her beliefs, as well as her encounter with the recanting Charles Darwin!

Since context is everything in history, the context of the Darwin family in Victorian times is paramount to this discussion. Darwin died a committed agnostic, but his beliefs were a source of tension between him and his wife, Emma. Emma Darwin was a devout Anglo-Unitarian. Darwin's writings and personal reflections were edited to remove some of his more controversial comments on religion.[15]

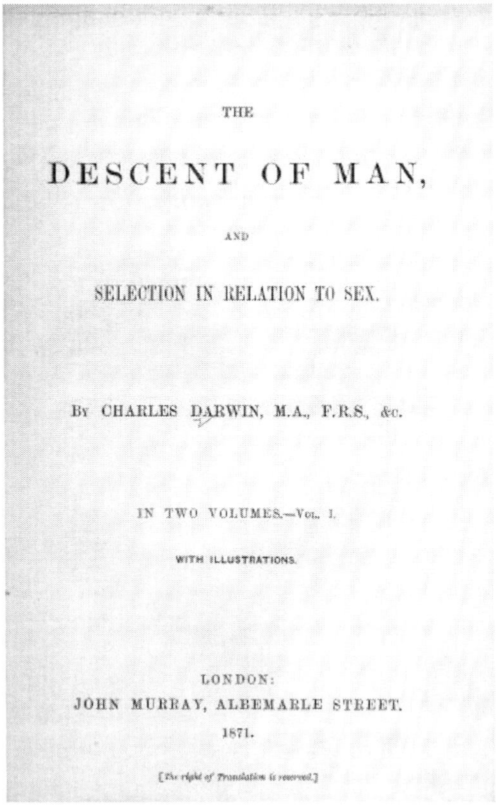
THE

DESCENT OF MAN,

AND

SELECTION IN RELATION TO SEX.

BY CHARLES DARWIN, M.A., F.R.S., &c.

IN TWO VOLUMES.—VOL. I.

WITH ILLUSTRATIONS.

LONDON:
JOHN MURRAY, ALBEMARLE STREET.
1871.

[The right of Translation is reserved.]

The Descent of Man (Vol.1), 1871, is the volume in which Darwin presents his "long argument" for human evolution. Wikimedia Commons. Darwin's earlier work, On the Origin of Species, did not deal with human evolution, or "transmutation," as it was then referred to. Darwin was correct is postulating that homo sapiens first appeared in Africa.

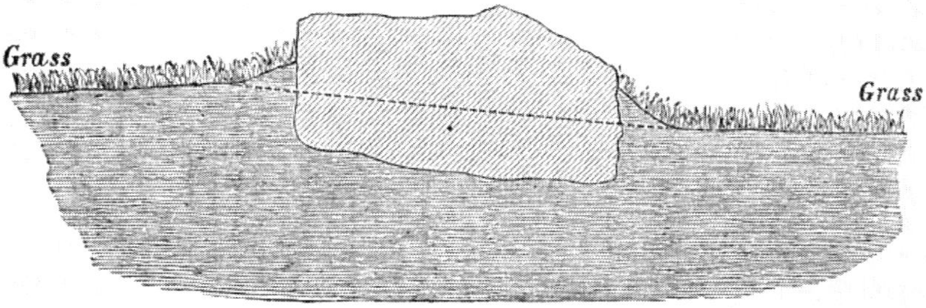

Darwin, *The Formation of the Vegetable Mould through the Action of Worms,* 1881. One of the Druid stones at Stonehenge. Darwin was trying to demonstrate how earthworms affect their environment. Wikimedia Commons. This book was published the year before Darwin died and contains more evidence refuting Lady Hope's claim.

One such example exemplifies this in his autobiography: "I can indeed hardly see how anyone ought to wish Christianity to be true; for if so the plain language of the text [the Gospels] seems to show that the men who do not believe, and this would include my Father, Brother, and almost all my best friends, will be everlastingly punished. And this is a damnable doctrine."[16] As the autobiography editor, Nora Barlow states, these words were left out of the *Life and Letters of Charles Darwin* which was first published within the decade after his death which included annotations by Emma Darwin.[17] Those present at Darwin's death were Emma, daughters Henrietta (Litchfield) and Bessy, and son Francis.[18] Henrietta was later quoted in the newspaper, *The Christian*, specifically refuting Hope's story as reported and promulgated in the United States.[19] Later, the publication of Darwin's thoughts on religion were disputed within the family, and, via a compromise, the more polemical statements were omitted and or edited.[20] Yet in none of these intra-familial arguments was there a discussion or argument over a deathbed confession/recantation. The sum and substance of the dispute center on Darwin's lack of faith in the orthodox beliefs, personal statements which his family viewed to be contrary to his Darwin's contemporaries within Victorian England. There is no mention of a conversion, a confession, or a recantation of his scientific accomplishments. If Darwin did find a newfound faith or a major change in his religious beliefs, the perfect place to add this would have been as an annotation in the autobiography, especially given the close proximity to his death when it was published and it would have still been fresh on the family's (i.e., Emma's) memory.

Since words and opinions attributed to Darwin were and are still used against him, we should also look at words we know he wrote, and right at the time of the alleged time of his interview with Lady Hope. In Darwin's final treatise on earthworms, he states the

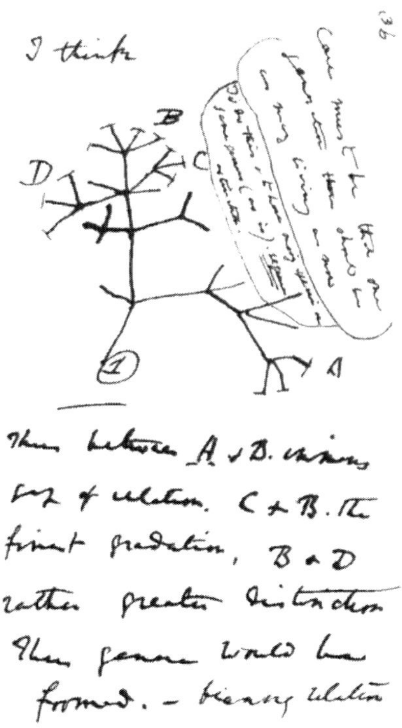

Charles Darwin, Tree of Life diagram, Wikimedia Commons. This well-illustration adorns artwork and can be seen on even tee shirts today.

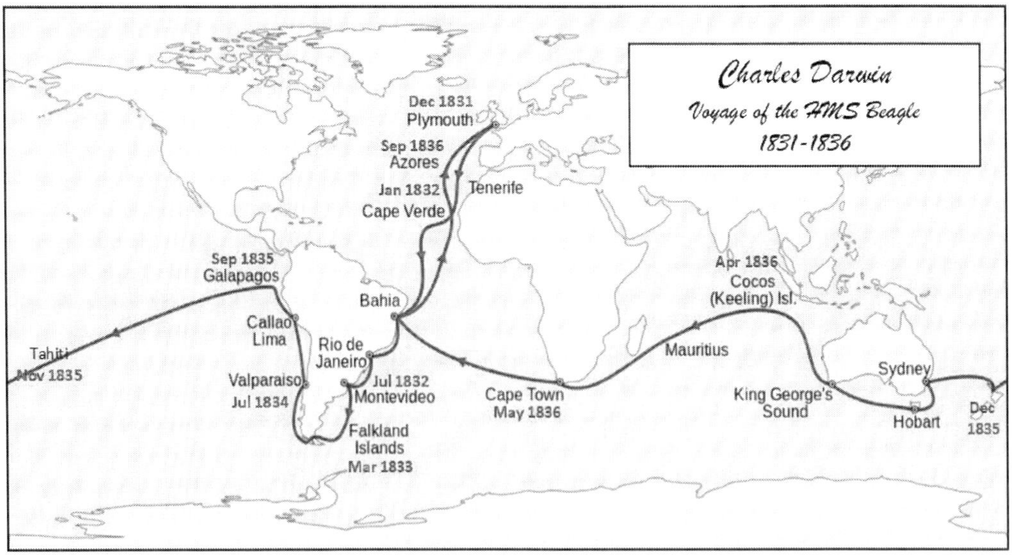

Voyage of the HMS Beagle 1831 – 1836. Wikimedia Commons augmented with additional dates by the editor. Charles Darwin's voyage aboard the *Beagle* lasted about five years. The publication of his book *Voyage of the Beagle,* which chronicled his voyage, brought him widespread acclaim and influenced his later published theories on evolution by natural selection."

following: "… but long before he [man] existed the land was in fact regularly ploughed, and still continues to be thus ploughed by earth-worms."[21] This closing statement shows Darwin did not accept the fundamentalist/literal claim of six days of creation as worms were here long before man, not "created" at the same time or on the same day, and that the worms shaped the land, not an all-powerful omniscient deity.[22]

Let us return to Lady Hope herself. Having worn out her welcome in her own country, she found a more welcoming, albeit gullible audience in America.[23] It is doubtful that Lady Hope had the clairvoyance to see where her story may go and be further exploited a century later. More likely, she tailored the story in the hopes of showing her influence on such a famous (or infamous) man like Charles Darwin would propel her to a second proselytizing career in America. Lady Hope did not intend to interfere or undermine the public teaching of evolution. Big things tend to have little beginnings. Lady Hope's story was either invented or embellished for her own self-importance. A story where the infamous Charles Darwin simply applauded her temperance work would have no audience; a story where a repentant "devil's chaplain" confessed to her would have a mass appeal. From a social-psychological standpoint, Lady Hope needed to belong to something or some group. And what better way to gain acceptance to a group than by telling that group a story you know they would accept uncritically? The idiom "never let the truth ruin a good story" comes to mind.

In the end, this demonstrates the importance of comparing reliable primary sources (e.g., Charles and Emma Darwin), and not relying on an unsubstantiated claim by an individual of dubious background. Lady Hope was fading into obscurity at the time she initially told her story in Massachusetts. Lady Hope was penniless and was looking to curry favor with new followers and supporters. In early twentieth-century American style, her claim went viral in fundamentalist Christian culture and different groups attempted to capitalize on it to discredit the teaching of evolutionary biology. The story even reached one of the most notable public personalities of his day, William Jennings Bryan, and that was prior to him being personally contacted by the beholders of the "Los Angeles Affidavit." Bryan probably saw the potential in the account, but nothing immediately useful to his crusade. It was not enough for the father to disown the child- all evidence of the child had to be impeached.

It was probably not Lady Hope's intention to attack the teaching of evolution, but her followers obviously appropriated the story for their own disreputable purposes. Since important details in the story change with different audiences, the accounts got tailored accordingly by those transmitting them in ways they deemed necessary. The specifics and overt references to evolution also get altered with each retelling. The Darwin family claiming they were present at Darwin's death doesn't totally destroy the fable as Hope's story doesn't claim their interview was his deathbed. Darwin's own words do contradict the recanting element implied/stated in the story, specifically in his monograph on earthworms and his autobiography. Henrietta's opinion that the story gained a threshold in America is spot on, even if her claim that she was with Darwin at his death is irrelevant given what Hope actually says. None of the writings by Darwin, whether in his autobiography, with familial annotations, or in his last scientific publication show a man changing his long-held views on the orthodox Anglican religion.

Bibliography

Bole, Simeon James. *The Battlefield of Faith*. University Park: College Press, 1940.

Bowler, Peter J. *Darwin Deleted: Imagining a World Without Darwin*. Chicago: The University of Chicago Press, 2013.

Browne, Janet. *Charles Darwin: Voyaging: A Biography*. Princeton: Princeton Univ. Press, 1995.

Browne, Janet. *Charles Darwin: The Power of Place*. Princeton: Princeton University Press, 2003.

Browne, Janet. *Darwin's Origin of Species: A Biography*. New York: Atlantic Monthly Press, 2006.

Bryan, William Jennings. "Archive Record." Nebraska State Historical Society Manuscript Collection. Accessed June 09, 2017. http://nebraskahistory.pastperfect online.com/archive/A8879A0B-3E29-4CEE-8E15-823224893044.

Caudill, Edward. *Darwinian Myths: The Legends and Misuses of a Theory*. Knoxville: University of Tennessee Press, 1999.

Clark, Ronald W. *The Survival of Charles Darwin: A Biography of a Man and an Idea*. New York: Random House, 1984.

Darwin, Charles. *Darwin on Humus and the Earthworms: The Formation of \Vegetable Mould through the Action of Worms; with Observations on their Habits*. Edited by Sir Albert Howard. 4th ed. London: Faber and Faber, 1966.

Darwin, Charles. *The Autobiography of Charles Darwin, 1809–1882*. Edited by Nora Barlow. New York: Norton, 1993.

Darwin, Charles. *From So Simple a Beginning: The Four Great Books of Charles Darwin*. Edited by Edward O. Wilson. New York: W.W. Norton & Co., 2006.

Darwin, Emma. *Emma Darwin, A Century of Family Letters, 1792–1896*. Edited by Henrietta Emma Darwin Litchfield. Vol. I-II. London: J. Murray, 1915. https://archive.org/stream/emmadarwincentur01litc#page/n7/mode/2up.

Desmond, Adrian, and James Moore. *Darwin: The Life of a Tormented Evolutionist*. New York: W.W. Norton & Company, 1994.

Keynes, Randal. *Annie's Box: Charles Darwin, his Daughter and Human Evolution*. London: Fourth Estate, 2002.

Larson, Edward J. *Summer for the Gods*. New York: Basic Books, 1997.

Larson, Edward J. *Evolution*. New York: Modern Library, 2003.

Levine, Lawrence W. *Defender of the Faith, William Jennings Bryan: the last decade, 1915-1925*. New York: Oxford University Press, 1968.

Litchfield, Henrietta. "Charles Darwin's Death-Bed. Story of Conversion Denied." *The Christian* (London), February 23, 1922. https://esc.illiad.oclc.org/illiad/illiad.dll?Action=10&Form=75&Value=8651.

Livingstone, David N., D. G. Hart, and Mark A. Noll, eds. *Evangelicals and Science in Historical Perspective*. New York: Oxford University Press, 1999.

Loy, James, and Kent M. Loy. *Emma Darwin: A Victorian Life*. Gainesville, FL: University Press of Florida, 2010.

Moore, James R. *The Darwin Legend*. Grand Rapids, MI: Baker Books, 1994.

Newman, Robert C. "The Darwin Conversion Story: An Update." Editorial. *Interdisciplinary Biblical Research Institute,* 1992. Accessed October 27, 2017. http://www.newmanlib.ibri.org/.

Numbers, Ronald L. *The Creationists*. Berkeley: University of California Press, 1993.

Phipps, William E. *Darwin's Religious Odyssey*. Harrisburg: Trinity Press International, 2002.

Quammen, David. *The Reluctant Mr. Darwin: An Intimate Portrait of Charles Dar-*

win and the Making of his Theory of Evolution. New York: Atlas Books/Norton, 2006.

Shermer, Michael. *Why Darwin Matters: The Case against Intelligent Design*. New York: Times Books, 2006.

"Wheaton's History." Wheaton. Accessed June 09, 2017. http://www.wheaton.edu/About-Wheaton/History.

Yates, Simon. "The Lady Hope Story: A Widespread Falsehood." The Lady Hope Story: A Widespread Falsehood. Accessed October 05, 2017. http://www.stephenjaygould.org/ctrl/ladyhope.html.

Notes

1 James R. Moore, *The Darwin Legend* (Grand Rapids: Baker Books, 1994), 24. This volume is unfortunately out of print. My work expands upon what Moore analyzed and his book contains many of the primary and secondary sources in question.

2 Moore, *Darwin Legend*, 81–82.

3 Moore, *Darwin Legend*, 89–90; J.W.C Fegan letter to J.A Kensit (of the Protestant Truth Society), May 1, 1925, in *The Darwin Legend* James R. Moore (Grand Rapids: Baker Books, 1994), 156, 180.

4 Moore, *Darwin Legend*, 22, 97; Feagen, letter to Kensit, 156; J.W.C Fegan letter to S. J. Pratt, May 22, 1925, in *The Darwin Legend*, James R. Moore (Grand Rapids: Baker Books, 1994), 162.

5 Moore, *Darwin Legend*, 91–93.

6 Moore, *Darwin Legend*, 94, 96.

7 Ronald L. Numbers, *The Creationists* (Berkley: University of California Press, 1993), 55; Lawrence W. Levine, *Defender of the Faith, William Jennings Bryan: The Last Decade, 1915-1925* (New York: Oxford University Press, 1968), 277.

8 Anette Parkinson Smith letter to William Jennings Bryan, June 7, 1922, in *The Darwin Legend*, James R. Moore (Grand Rapids: Baker Books, 1994), 123–26.

9 Smith, letter to Bryan, 123–26; Melville Dozier et al., affidavit to William Jennings Bryan, in *The Darwin Legend*, James R. Moore (Grand Rapids: Baker Books, 1994), 127.

10 Smith, letter to Bryan, Moore, 126.

11 Moore, *Darwin Legend*, 127–28. This author's tenth-grade biology teacher shared a similar view as William Jennings Bryan did. This author once approached the afore-

mentioned teacher, Mr. Barrow, about the claim that Darwin renounced his theory of evolution. Mr. Barrow's response did not betray whether he himself believed this story, but he simply stated that this was irrelevant; the evidence that Darwin found, and by implication, those that succeeded him also discovered, supports the theory of evolution by Natural Selection. Another high school teacher of mine, Mr. Lenhart was probably the first person to teach this author the difference between objective (scientific) source reliable sources and apologetics.

12 Moore, *Darwin Legend*, 124–25.

13 Simeon James Bole, *The Battlefield of Faith* (University Park: College Press, 1940), Preface; Moore, *Darwin Legend*, 130; William Jennings Bryan "Archive Record." Nebraska State Historical Society Manuscript Collection, http://nebraskahistory.pastperfectonline.com/archive/A8879A0B-3E29-4CEE-8E15-823224893044.

The summary states that Bryan requested Bole, then at Wheaton College, to appear as an expert witness at the Scopes Monkey Trial; Wheaton's History. Given the liberal arts history of this institution, it is no wonder why Bole felt "persecuted" and left for John Fletcher College.

14 Bole, *Battlefield of Faith*, 166–69.

15 Moore, *Darwin Legend*, 26, 36.

16 Charles Darwin, *The Autobiography of Charles Darwin, 1809-1992*. Edited by Nora Barlow (New York: Norton, 1993), 72.

17 Darwin, *Autobiography of Charles Darwin*, Preface, 72–73.

18 Moore, *Darwin Legend*, 52.

19 Henrietta Litchfield, "Charles Darwin's Death-Bed. Story of Conversion Denied," *The Christian*, (London), February 23, 1922, https://esc.illiad.oclc.org/illiad/illiad.dll?Action=10&Form=75&Value=8651; Emma Darwin, *Emma Darwin, A Century of Family Letters, 1792-1896. Edited by Henrietta Litchfield, Vol. I-II* (London: John Murray, 1915), 251–54. https://archive.org/stream/emmadarwincentur01litc#page/n7/mode/2up.

There is no mention of any religious conversion or transformation, just the memories Emma cherished with her husband. There is also no references to Hope or any female evangelist.

20 Moore, *Darwin Legend*, 61–63.

21 Charles Darwin, *The Formation of Vegetable Mould through the Action of Worms with Observations on their Habits*. (London: Faber and Faber, 1966), 148.

22 This does not account for the conflicting versions of exactly, and in what order, mankind was created throughout Chapters 1 and 2 of Genesis.

23 Moore, *Darwin Legend*, 94.

Balloons of the Civil War: The Birth of Multi-Dimensional Warfare

Michael G. Stroud
American Military University

Abstract

Giant, colorful balloons rising gently into the blue sky is the modern image of balloons. Few realize that these lighter-than-air balloons and their passionate aeronauts were put to military use by both sides during the tumultuous American Civil War to turn the tide in their favor. The forward-thinking application of the giants of the sky changed warfare forever. This study examines the military application of balloons by both the Union and the Confederacy, as well as their daring aeronauts who laid the foundations for multi-dimensional warfare. The use of diverse sources reveal the history of military ballooning, its use in the American Civil War, and identify the wars' most prominent aeronauts. This study concludes with the legacy of ballooning in the Civil War, and how aeronauts and their actions laid the foundations for multi-dimensional warfare in the future.

Keywords: Balloon Corps, Aeronauts, Civil War, Ascensions, Hydrogen Gas Generator, Peninsula Campaign, Silk Dress Balloon, Geography, Tethered, Airtight

Globos de la Guerra Civil: el nacimiento de la guerra multidimensional

Resumen

Globos gigantes y coloridos que se elevan suavemente hacia el cielo azul es la imagen moderna de los globos. Pocos se dan cuenta de que estos globos más livianos que el aire y sus apasionados aeronautas fueron utilizados militarmente por ambos bandos durante la tumultuosa Guerra Civil estadounidense para cambiar el rumbo a su favor. La aplicación con visión de futuro de los gigantes del cielo cambió la guerra para siempre. Este estudio examina la aplicación militar de los globos tanto por parte de la Unión como de la

doi: 10.18278/sshj.10.3.4

Confederación, así como de sus atrevidos aeronautas que sentaron las bases para la guerra multidimensional. El uso de diversas fuentes revela la historia de los globos militares, su uso en la Guerra Civil estadounidense e identifica a los aeronautas más destacados de la guerra. Este estudio concluye con el legado de los globos aerostáticos en la Guerra Civil y cómo los aeronautas y sus acciones sentaron las bases para la guerra multidimensional en el futuro.

Palabras clave: Cuerpo de Globos, Aeronautas, Guerra Civil, Ascensiones, Generador de Gas Hidrógeno, Campaña Península, Globo Vestido de Seda, Geografía, Atado, Hermético

美国内战中的热气球：多维度战争的诞生

摘要

巨大的彩色热气球缓慢升入蓝天，这是热气球的现代形象。很少有人意识到，这些比空气还轻的热气球及其狂热的驾驶员曾在动荡的美国内战期间被合众国与联盟国用于军事，以期扭转战局。对热气球的前瞻性应用永远改变了战争。本研究分析了合众国与联盟国对热气球的军事应用，并分析了那些为多维度战争打下基础的勇敢的热气球驾驶员。使用源自不同来源的数据，揭示了军事热气球史、其在美国内战中的使用，并识别了内战中最杰出的驾驶员。本研究的结论包括热气球在美国内战中的影响，以及驾驶员及其行动如何为未来的多维度战争奠定基础。

关键词：热气球军团，热气球驾驶员，美国内战，上升，氢气发生器，半岛战役，丝绸热气球，地理，系绳，密封

The American Civil War (1861-1865) was a vastly unique conflict that extended far beyond brother against brother and families torn apart based on ideology. The conflict spurred many unique inventions and evolving technologies, which came into play for the first time in America, such as the *CSS Hunley* submarine—the first of its kind to sink an enemy vessel—or the important use of the telegraph for both battlefield communications and civilian newsmen to reporting on the war itself. It was another innovation, however, which while not new in its creation and use elsewhere, was first put to military use in the United States. Ballooning, under the passionate advocating of key aeronauts (someone who travels in a balloon), was utilized by both sides in the Civil War, thus paving the way for multi-dimensional warfare.

History of Ballooning in Warfare

The balloon was first created in 1783 by the French Montgolfier brothers, Joseph-Michel (1740-1810) and Jacques-Étienne (1745-1799), who utilized hot air to rise almost 6,000 feet in a 35-foot diameter balloon for the first time.[1] This demonstration was witnessed by Benjamin Franklin (1706-1790), who was in Paris at the time as part of the American delegation signing the Treaty of Paris to formally end the American Revolution. Franklin was so impressed with the demonstration and its potential that he had remarked: "[F]ive thousand balloons, capable of raising two men each could not cost more than five ships of the line, and where is the prince who can afford so to cover his country with troops for its defense as that ten thousand men descending from the clouds might not in many places do an infinite deal of mischief before a force could be brought together to repel them?"[2]

Franklin was not alone in grasping the military potential of balloons, as various European countries seized on the opportunities presented by this invention. The French put balloons to work during the Revolutionary Wars with the creation of the Aerostatic Corps, which proved its worth at the Battle of Fleurus on June 26, 1794. Its observations of enemy positions and movements at the battle allowed the French to deliver a decisive defeat of the Allied army of the First Coalition. Napoleon Bonaparte (1769-1821) had balloons with him during his arrival in Egypt in 1798, but the British fleet destroyed them before they could be unloaded and brought into use. Even later, the future Emperor had proposed using balloons as part of his invasion plans of the British Isles, which never came to fruition.

The British, Russians, Dutch, Austrians, and others utilized balloons in either military planning such as the British in towing balloons to spy on the French, or the Russians who wanted to drop bombs on Napoleon during his 1812 invasion. The aerial reconnaissance value of balloons was realized in the Franco-Austrian War of 1859, which allowed for a French victory, and later in Franco-Prussian War in 1870 when Napoleon III (1808-1873) contemplated using an observation balloon as a means of escape from the Prussian army—he eventually surrendered to the Prussians at Sedan on 3 September 1870.

Other military applications for the use of balloons included as a bomber, when in 1849 during the Siege of Venice, the Austrians deployed balloons of under 20 feet in diameter to drop explosives on the city. Their unpredictability in shifting winds led to their eventual disuse in this capacity when the balloons and bombs began drifting back over the Austrians themselves. Aerial photography and its strategic and tactical value were also introduced near Paris in 1858, which played an import-

ASCENT OF THE 19TH SEPTEMBER, 1783, AT VERSAILLES.

"Ascent of the 19th September, at Versailles, 1783," [or Later] Photograph, Library of Congress, accessed December 16, 2021. https://www.loc.gov/item/2007681716/.

"Prof. T.S.C. Lowe, Civil War balloonist," [Between 1861 and 1865] Photograph, Library of Congress, accessed December 16, 2021. https://www.loc.gov/item/2002721619/.

Brady, Matthew B., "Professor Lowe's military balloon near Gaines Mill, Virginia," Photograph, Library of Congress, accessed December 16, 2021. https://www.loc.gov/item/2012649020/.

ant role in the previously mentioned Franco-Austrian War.

The United States was not isolated from the military potential of balloons pre-Civil War; however, the strategy was never put into practical use or fully realized. The Second Seminole War of 1835-1842 between the native Seminoles and the Federal government was one instance in which it was suggested by Colonel John Sherburne that balloons could ascend and ascertain the position of the Seminoles by the light of their campfires at night.[3] The idea was strongly considered, but eventually rejected based on the terrain of the campaign area.

Later, in the Mexican-American War (1846-1848), American troops were faced with various Mexican strongholds and fortresses that looked to be impregnable from traditional assaults. In this environment, the launch a large, tethered balloon carrying over 18,000 pounds of high explosives to bomb said targets into submission was proposed. This proposal was either never received or ignored as no response had been made and subsequently no action taken to implement this bold tactic.

Military ballooning finally gained a foothold on 17 June 1861 when balloon pioneer and inventor, Thaddeus S.C. Lowe (1832-1913) gave a demonstration in Washington D.C. with President Lincoln in attendance. Rising over 500 feet above the ground in his tethered balloon the *Enterprise*, Lowe telegraphed back to the President what he saw from his vantage. "This point of observation commands an area nearly 50 miles in diameter. The city, with its girdle of encampments, presents a superb scene."[4] The feat was not lost on Lincoln, who supported the creation of the U.S. Balloon Corps with Lowe at its head. Military balloons now had a foothold in the United States.

Prominent Aeronauts of the Civil War

Thaddeus Lowe could easily be credited as the father of America's aeronautics program and therefore, its most prominent aeronaut. A native of New Hampshire, Lowe demonstrated early, passionate interests in both science and aeronautics, which led him to make his first balloon ascension at the age of twenty-six in 1858.[5] In the years following this initial endeavor, Lowe set about securing financial backing for a balloon trip across the Atlantic that he was confident was possible. At the behest of friends, including the noted scientist and First Secretary of the Smithsonian, Joseph Henry (1797-1878), Lowe proceeded with further land-based balloon travel experiments before attempting his dangerous Atlantic trip. The firing on Ft. Sumter, on 12 April 1861, presented an opportunity for Lowe to offer the Federal government the use and benefit of his ballooning services in the coming war. Through a combination of political connections, charisma, determination, and the approval of President Lincoln, Lowe was placed at the head of the newly created U.S. Balloon Corps in July of 1861.

Though the most visible and well known of the Union's aeronauts, Lowe was not alone. Rhode Islander James Allen (1824-1897) was also well known at the time for his numerous balloon ascensions, beginning in 1856; with the outbreak of hostilities, he packed up his balloons and gear and made way with the First Rhode Island Regiment to Washington, D.C. Once there, Allen's demonstration of his balloons ended in their destruction—one through impalement due to strong winds and the other through a coal accident during inflation.[6] Allen went on to serve under Lowe as part of the Balloon Corps.

Another early aeronaut was Pennsylvania native John Wise (1808-1879). Having begun his ballooning experience at the age of 27, Wise accumulated hundreds of successful ascensions by the outbreak of war in 1861, even having earned the reputation as "the aeronaut of his time."[7] This reputation led to his direct appointment as a balloonist by the Chief of Topographical Engineers, Maj. Hartman Bache (1798-1872). After having secured funding from the military to build a 20,000 cubic foot balloon, completed on 16 July 1861, Wise encountered several problems in his attempt to put the balloon in the field. After the balloon's destruction from being caught in some trees, and then a follow up effort resulting in the balloon being blown away, Wise was reprimanded and his time as a military aeronaut was over. He served out the war as a cavalry officer with the Pennsylvania Volunteers.

A final prominent Union balloonist was John La Mountain (1830-1878), a compatriot of John Wise. The New York native had made a half a dozen ascensions with Wise before venturing out on his own, when he gained notoriety for his 800-mile trip from St. Louis to Henderson, New York, aboard his 100,000+ cubic foot balloon, the *Atlantic.*[8] Having his offers of service initially rebuffed, La Mountain was given an opportunity in late July of 1861 to support Fortress Monroe in Virginia. His successful operations there lead to eventual testing of free ascension (no tethered lines) and observation of Confederate positions in September 1861. Having successfully endured accidental friendly fire, La Mountain became embroiled in a heated back-and-forth with Thaddeus Lowe to secure use of one of Lowe's balloons, which culminated with La Mountain's dismissal from military service by General George B. McClellan (1826-1885). The Union military had decidedly backed Lowe and his Balloon Corps as the way forward.

The Confederacy, due to extremely limited resources, never fielded an effective or comparable balloon program. It was because of these deficiencies, that there is little in the form of Confederate aeronauts—save one. Lt. Col. Edward Porter Alexander (1835-1910), formerly of West Point, was appointed as the commander of the Confederacy's balloon program in Virginia. The Confederates would make use of a few balloons under Alexander to gather intelligence on Union troops, but their construction materials, made from dresses bought in Charleston, S.C., combined with their only source of gas being in Richmond, amounted to an impractical expense for their finite resources.[9] The Confederacy's military use of balloons had effectively end by early 1863.

Military Use of Balloons in the Civil War

Thaddeus Lowe, ensconced in his new role as "commander" of the newly created U.S. Balloon Corps, quickly set about "militarizing" his balloon corps for the tasks at hand. Utilizing double layers of silk with multiple coats of varnish, Lowe greatly improved both the strength and airtight capacity of what soon became a force of seven balloons. To help prevent the accidental severing of tethered ascensions, Lowe and his team requisitioned and put into place thick and strong 5,000-foot-long ropes that could safely anchor aeronauts in various weather conditions. Additionally, due to the varying needs of the military, Lowe designed and constructed balloons of varying sizes, with the largest being able to accommodate up to five men. Finally, timely communication from the observations of the aeronauts to the officers on the ground was key and therefore, Lowe was able to secure the use of telegraph equipment with eventually an entire unit along with operators being assigned to the Balloon Corps.[10]

To further improve the military viability of his balloons, and realizing that reliable transportation to and from the coal gas works of cities was needed to inflate their balloons, Lowe developed a portable hydrogen gas generator. This now allowed for balloon operations outside of direct city limits and roads, though the heavy metal tanks were still extremely cumbersome and difficult for their wagons and horses. Soon, the Balloon Corps acquired and converted a barge for rapid transportation along the Potomac River, as well as for conducting balloon operations from its deck, effectively making it the first aircraft carrier.

Lowe and his fellow aeronauts were considered civilian employees, with Lowe making the daily sum of $10 and his team making half that amount (or less). This lack of a military commission presented numerous problems for Lowe and his team, not least of which was the securing and training of the needed manpower for his balloon operations.[11] Lowe engaged in constant negotiations with nearby regimental commanders from wherever they were to secure and train the 30-man teams needed to operate the fleet of balloons. By the war's end, over 17 regiments had provided the manpower for the Balloon Corps to conduct operations.

The U.S. Balloon Corps served in several battles and campaigns, but its greatest military contributions were after the Battle of First Bull Run on 21 July 1861 and then again during the Peninsula Campaign of spring and summer of 1862. The devastating Union loss at First Bull Run, a mere 30 miles southwest of Washington D.C., invoked panic in civilians, politicians, and the military alike that the Confederates would move on the capital. To properly assess the situation, Lowe boarded the *Enterprise* on 24 July and ascended skyward. Seeing no evidence of Confederate forces coalescing for an attack on Washington D.C., Lowe's report alleviated the fears of all in the area and elevated the worth of the Balloon Corps.

Following this operational success, Lowe, as Chief of Aeronautics, was tasked by McClellan with deploying a coordinated fleet of four balloons along the Potomac River to serve as lookouts for further Confederate troop movements and to alert the Union as to any surprise attacks. This commission was also utilized to carry draftsmen and the like upward, so that detailed maps of the terrain, enemy fortifications, and troops positions could be made and relayed to Union officers. The success of the Balloon Corps during this time is considered minimal as they only verified that no Confederate forces were massing to strike the Capital, and outright incorrect by others who cited that Lowe's reports of Confederate guns were inaccurate.[12]

The Balloon Corps played a larger role in Gen. McClellan's Peninsula Campaign in the spring and summer of 1862. Along with near the entirety of the Corps to join McClellan's campaign, Lowe and his team were ordered to observe Confederate defenses around Yorktown in anticipation of a major siege operation. Several Union officers, led by the balloonist Brig. Gen. Fitz John Porter (1822-1901) and including Capt. George A. Custer (1839-1876) made numerous ascensions to carefully take note of enemy fortifications and strength. Once Lowe observed that the Confederates had fewer troops than originally believed and had subsequently abandoned Yorktown, he remarked that these observations and drawings by officers aboard his balloons "were of [the] greatest importance, and readily enabled the commanding officer to decide what course he would pursue."[13]

These balloon flights to assess Confederate defenses around Yorktown found the aeronauts coming under Confederate attack. Rifle and artillery fire were utilized in their efforts, which fell far short of their targets, yet made history as the first use of "antiaircraft fire" in American history. Yorktown commander Confederate General John B. Magruder (1807-1871) took the constant flights of the Union's balloons so seriously that he attempted to counter-attack them by sending up a cotton bag filled with hot air.[14] These efforts ended disastrously with no effect.

Lowe and his men continued to serve at McClellan's pleasure during the Peninsula Campaign, providing reports of enemy troop movements, trains, wagons, and terrain to assist in the campaign. From Far Oaks to Gaines' Mill to Richmond, Lowe and his Balloon Corps provided intelligence to McClellan and the Union that was considered marginally helpful, spotty, and suspect in its accuracy.

During the Peninsula Campaign, the Confederacy's balloon efforts finally took off, specifically during the defense of Yorktown. The 21-year-old Captain John Randolph Bryan (1806-1887), serving under Magruder, made multiple ascensions that allowed him to draw sketches depicting Union infantry, cavalry, artillery, wagon trains, and important geographical features such as streams and roads.[15] Bryan's only night ascent ended in a harrowing adventure for the Captain, in which he crash landed, then stealing a horse to report his observations to his officers; the balloon itself was lost and never seen again.

A Confederate replacement, constructed by Captain Langdon Cheves (1814-1863) of South Carolina, was put into use during the Seven Days' Campaign of late June, early July 1862. Measuring over 24 feet in diameter and made of reinforced silk material, the "Silk Dress Balloon" was assembled in Savannah, GA, inflated with hydrogen gas in Richmond, and put into service making several ascensions.[16] The Silk Dress Balloon was captured by Union naval vessels on 4 July 1862, as it was being transported by the Confederate tug *Teaser* on the James River. Another silk balloon was constructed and put into limited use in the summer of 1863, but it proved a casualty of the siege of Charleston, thus ending the Confederacy's balloon operations. Their inability to generate hydrogen gas in the field, coupled with its generally high cost, and manpower requirements made it impractical for the meager resources of the Confederate States.

Legacy of the Aeronauts

Aeronauts such as Thaddeus Lowe were passionate about their craft and believed wholly that balloons could revolutionize warfare and in turn, the wider world. Their direct impact on the Civil War proved to be minimal and is often relegated to a mere footnote. It is the concept of air dominance and its many benefits that is its true legacy. The force could see farther than their enemy, rain bombs from above, and transport men through the air; it could potentially be supreme. The introduction of balloons in a military application were the first steps toward such dominance.

On the Union side, Lowe and his Balloon Corps were shut down in 1863, as it was determined to be too high of a cost both financially and in personnel. No matter how persuasive or committed Lowe and his team were, they could not overcome the deficits of the technology of the time either, fighting a losing battle with the strong unpredictable winds of Mother Nature that oftentimes made ballooning precarious.

In the South, the matter was entirely resource related. The Confederacy was at a distinct and clear disadvantage in resources at the very outbreak of hostilities and these only worsened as the war wore on, greatly impeding any potential for a sustained Confederacy Air Corps. Though some efforts were made, such as with the Silk Dress Balloon, the Confederacy never had the resources that the Union had available to fund and outfit a proper balloon unit. General Longstreet (1821-1904) later lamented this when he said: "The Federals had been using balloons in examining our positions, and we watched with anxious eyes their beautiful observations as they floated high up in the air, well out of range of our guns."[17]

The legacy of the aeronauts of the Civil War is not in their decisive impact on the war itself, as they did no such thing. Rather it is in their passion for taking to the sky and bringing to bear various elements, from communications to photography to bombing, to add another dimension to warfare and therefore usher in the beginning of multi-dimensional combat.

Notes

1 Steven D. Culpepper, "Balloons of the Civil War," (Masters diss., U.S. Army Command and General Staff College, 1994), 5, https://apps.dtic.mil/sti/pdfs/ADA284682.pdf.

2 Ben Fanton, "View from above the battlefield," *America's Civil War*, September 2001, https://www.proquest.com/magazines/view-above-battlefield/docview/223340644/se-2?accountid=8289.

3 Ibid.

4 Eugene H. Baker, *The Civil War, Forward to Richmond* (New York, NY: Time Inc., 1983), 147.

5 Duane J. Squires, "Aeronautics in the Civil War," *The American Historical Review* 42, no. 4 (1937): 655. https://doi.org/10.2307/1839448.

6 Steven D. Culpepper, "Balloons of the Civil War," (Masters diss., U.S. Army Command and General Staff College, 1994), 11, https://apps.dtic.mil/sti/pdfs/ADA284682.pdf.

7 Ibid., 12.

8 Ibid., 13.

9 James Scythes and Spencer C. Tucker, "Balloons: American Civil War," in *World at War: Understanding Conflict and Society,* 2021, accessed December 15, 2021, https://worldatwar2-abc-clio-com.ezproxy2.apus.edu/Search/Display/1067123.

10 Joseph C. Scott, "The Infernal Balloon: Union Aeronautics During the American Civil War," *Army History*, no. 93 (2014): 9, https://search-ebscohost-com.ezproxy2.apus.edu/login.aspx?direct=true&AuthType=ip&db=tsh&AN=98586024&site=ehost-live&scope=site.

11 Ben Fanton, "View from above the battlefield," *America's Civil War*, September 2001, https://www.proquest.com/magazines/view-above-battlefield/docview/223340644/se-2?accountid=8289.

12 Joseph C. Scott, "The Infernal Balloon: Union Aeronautics During the American Civil War," *Army History*, no. 93 (2014): 11, https://search-ebscohost-com.ezproxy2.apus.edu/login.aspx?direct=true&AuthType=ip&db=tsh&AN=98586024&site=ehost-live&scope=site.

13 Ibid., 13.

14 Duane J. Squires, "Aeronautics in the Civil War," *The American Historical Review* 42, no. 4 (1937): 663, https://doi.org/10.2307/1839448.

15 Steven D. Culpepper, "Balloons of the Civil War," (Masters diss., U.S. Army Command and General Staff College, 1994), 23, https://apps.dtic.mil/sti/pdfs/ADA284682.pdf.

16 Ben Fanton, "View from above the battlefield," *America's Civil War*, September 2001, https://www.proquest.com/magazines/view-above-battlefield/docview/223340644/se-2?accountid=8289.

17 Steven D. Culpepper, "Balloons of the Civil War," (Masters diss., U.S. Army Command and General Staff College, 1994), 24-25, https://apps.dtic.mil/sti/pdfs/ADA284682.pdf.

Bibliography

Baker, Eugene H. *The Civil War Forward to Richmond.* New York, NY: Time Inc., 1983.

Culpepper, Steven D. "Balloons of the Civil War." Masters diss., U.S. Army Command and General Staff College, 1994. https://apps.dtic.mil/sti/pdfs/ADA284682.pdf.

Fanton, Ben. "View from above the battlefield." *America's Civil War*, September 2001. https://www.proquest.com/magazines/view-above-battlefield/docview/223340644/se-2?accountid=8289.

Scott, Joseph C. "The Infernal Balloon: Union Aeronautics During the American Civil War." *Army History*, no. 93 (2014): 6-27. https://search-ebscohost-com.ezproxy2.apus.edu/login.aspx?direct=true&AuthType=ip&db=tsh&AN=98586024&site=ehost-live&scope=site.

Scythes, James, and Spencer C. Tucker. "Balloons: American Civil War." In *World at War: Understanding Conflict and Society*, 2021. Accessed December 15, 2021. https://worldatwar2-abc-clio-com.ezproxy2.apus.edu/Search/Display/1067123.

Squires, Duane J. "Aeronautics in the Civil War." *The American Historical Review* 42, no. 4 (1937): 652-69. https://doi.org/10.2307/1839448.

Media Sources

"Ascent of the 19th September, at Versailles, 1783 [or Later]." Photograph. Library of Congress. Accessed December 16, 2021. https://www.loc.gov/item/2007681716/.

Brady, Matthew B. "Professor Lowe's military balloon near Gaines Mill, Virginia."

Photograph. Library of Congress. Accessed December 16, 2021. https://www.loc.gov/item/2012649020/.

"Prof. T.S.C. Lowe, Civil War balloonist [Between 1861 and 1865]." Photograph. Library of Congress. Accessed December 16, 2021. https://www.loc.gov/item/2002721619/.

An Intersection of Myth and Memory: Hollywood Meets the Sullivan Brothers

Marisa Kahla
American Military University

ABSTRACT

In December 1942, Aletta Sullivan received the worst news of her life. The ship her sons had been serving on, the USS *Juneau*, had been destroyed during the naval battle of Guadalcanal and all five of her sons, the "fighting Sullivans" as they had been affectionately dubbed by the press, were missing in action and feared dead. The Sullivan brothers, with their boyish charms and wholesome Midwestern upbringing, became overnight sensations when they all enlisted together following the attack on Pearl Harbor and insisted on all serving on the same ship. The boys became media darlings, and their image was used for a variety of promotional material for enlistment and the growing war effort on the home front. Their deaths, however, would gain them mythical status, their story transformed from one of tragedy into triumph. The Sullivan brothers became the literal poster boys for Hollywood and media mythmaking at its finest, and their lives and sacrifices were used to effectively shape the way the world viewed not only the war itself but wartime sacrifice as a whole. The Sullivans made history, but it was Hollywood that made the Sullivans.

Keywords: Sullivan brothers, USS *Juneau*, Battle of Guadalcanal, naval battle, soldier death, grief and loss, film and media, war production, propaganda, Office of War Information

Una intersección de mito y memoria: Hollywood conoce a los hermanos Sullivan

RESUMEN

En diciembre de 1942, Aletta Sullivan recibió la peor noticia de su vida. El barco en el que habían estado sirviendo sus hijos, el USS Juneau, había sido destruido durante la batalla naval de Guadalcanal y sus cinco hijos, los "Sullivan luchadores", como los había apodado cariñosamente la prensa, estaban desaparecidos en combate y temidos. muerto. Los hermanos Sullivan, con sus encantos

doi: 10.18278/sshj.10.3.5

juveniles y su sana educación del Medio Oeste, se convirtieron en sensaciones de la noche a la mañana cuando todos se alistaron juntos después del ataque a Pearl Harbor e insistieron en que todos sirvieran en el mismo barco. Los muchachos se convirtieron en los favoritos de los medios y su imagen se usó para una variedad de material promocional para el alistamiento y el creciente esfuerzo de guerra en el frente interno. Sin embargo, sus muertes les otorgarían un estatus mítico, su historia se transformaría de una tragedia a un triunfo. Los hermanos Sullivan se convirtieron literalmente en los muchachos del cartel de Hollywood y la creación de mitos mediáticos en su máxima expresión, y sus vidas y sacrificios se utilizaron para dar forma efectiva a la forma en que el mundo veía no solo la guerra en sí, sino el sacrificio en tiempo de guerra en su conjunto. Los Sullivan hicieron historia, pero fue Hollywood quien hizo a los Sullivan.

Palabras clave: hermanos Sullivan, USS Juneau, Batalla de Guadalcanal, batalla naval, muerte de soldados, duelo y pérdida, cine y medios, producción de guerra, propaganda, Oficina de Información de Guerra

传闻与记忆的交叉：好莱坞遇上沙利文兄弟

摘要

1942年12月，阿莱塔·沙利文收到了一生中最坏的消息。她的五个儿子所服务的朱诺号轻巡洋舰在瓜达尔卡纳尔岛战役中被摧毁，并且她的五个儿子（此前被媒体亲切地称为“战斗的沙利文”）在行动中失踪，恐怕已经阵亡。沙利文五兄弟凭借男孩魅力和良好的美国中西部成长经历，曾因“其在珍珠港袭击事件后全部入伍并坚持在同一艘船上服务一事”而在一夜间引起轰动。这五兄弟成为媒体宠儿，他们的形象被用于一系列入伍宣传材料和越来越多的大后方准备工作。不过，他们的阵亡为其带来了虚构的身份，他们的故事从悲剧转变为胜利。沙利文五兄弟成为了好莱坞和媒体传奇制造的最佳海报男孩，他们的生活和牺牲被用于有效影响全世界如何看待这场战役，以及从整体上看待战争牺牲。沙利文五兄弟创造了历史，但成就他们的是好莱坞。

关键词：沙利文兄弟，朱诺号，瓜达尔卡纳尔岛战役，海战，士兵死亡，悲痛和损失，电影和媒体，战争片，宣传，美国战争情报局

The twisted hull of the ship rests stretched out across the ocean floor, her algae-covered turrets illuminated in the murky grey-green gloom of the water.[1] The siding is warped and jagged, mortal wounds caused by an enemy torpedo in the heat of battle. This has been the resting place of the USS *Juneau* for over seventy-five years—her name and legacy linked to unimaginable tragedy. During the naval battle of Guadalcanal, the *Juneau* was split in half by a Japanese torpedo; the bisected ship sank in a matter of seconds, taking nearly all hands with it. Among the dead and missing were the "fighting Sullivans," five brothers from Waterloo, Iowa, who had gained national attention the year before when they all enlisted together. Their deaths shook the entire nation and an outpouring of sympathy showered the boy's grieving parents, who had endured "the greatest single blow suffered by any one family . . . in American naval history."[2]

In the months and years that followed, the Sullivan brothers became the literal poster boys for the war effort on the home front—their names and images used to promote a variety of political, social, and commercial causes. They helped shape public opinion and American culture, their deaths becoming emblematic of courage, duty, and sacrifice. However, the story that captured the nation's attention and its sympathy was only partially true; in reality, the Sullivan brothers served as a perfect example of Hollywood and military myth-making at its finest. The uglier parts of the story were primed and polished and sometimes even completely ignored in favor of presenting a neat and clean version of reality to the public. The Sullivans made history, that much is certain, but it was really Hollywood that made the Sullivans.

Fame and publicity were the furthest things from anyone's mind as the Sullivan family bustled around their little house on December 7, 1941. It was a chilly, overcast day with the threat of snow looming on the horizon.[3] The family patriarch, Tom Sullivan, was upstairs resting, early mornings from his job at the railway taking their toll. Downstairs, Aletta Sullivan and her mother, May Abel, were busy tidying up the house and cooking dinner.[4] Soon enough the house would be filled with the couple's six children: five sons, George, Frank, Madison, Joseph, and Albert, and their only daughter, Genevieve. Sunday dinner was an important tradition in the Sullivan household and the children, while fully grown, were still regular attendants. It was a time for them to get together and enjoy good food and good company within the safe little walls of their family home.

Across town, the three older Sullivan boys, George, Frank, and Joseph, had spent most of the day at the Black Hawk motorcycle club. The 1920s had introduced a new souped-up version of motorized bicycles to the United States and by the 1930s, Harley Davidson held the monopoly on the market.[5] The boys had spent the day relaxing and discussing motorcycles, eventually pulling themselves away later that afternoon to head back home. Paul Hamilton, a friend of the brothers, had been invit ed to join them for dinner, so the group left together to meet up with the rest of the Sullivan family. However, instead of arriving to a warm welcome, the young men were met with panic.

Aletta, who had been listening to the radio while she cooked, came running out of the house to meet her boys, anxiously exclaiming that the Japanese had just attacked Pearl Harbor.[6] Within minutes the Sullivan family and their guests were huddled around the radio, listening in silent horror as the reports were read. The two younger Sullivan brothers, Madison and Albert, along with Genevieve, arrived at the house shortly thereafter and joined the rest of the family in the living room. As the damage reports and the names of those missing and killed during the attack were read, the Sullivan children were dismayed to find that they now had a personal connection to the tragedy. Bill Ball, a friend and fellow sailor of George and Frank—and a romantic acquaintance of Genevieve—had perished when the USS *Arizona* sank.[7] If the attack had been the catalyst, the loss of Bill and the desire to avenge him was what galvanized the Sullivan boys.

The brothers knew there would be an immediate need for sailors and soldiers—that every able-bodied young man in the country would be expected to rise up and defend their homeland. Talk in the Sullivan household immediately shifted to discussions about the Navy and how soon they could enlist. Pearl Harbor had given the Sullivan brothers a new direction and purpose in life. George Sullivan, ever the leader and the voice of his younger siblings, addressed them all soberly: "Well, I guess our minds are made up, aren't they fellows? And when we go in, we want to go in together. If the worst comes to worst, why, we'll all have gone down together."[8] No one could know how prophetic that statement would become in the months to follow.

On Friday, December 26, the Sullivan brothers joined the Navy *en masse* at the recruitment station, which had once served as the Waterloo Post Office.[9] They certainly were not the first to enlist following the attack on Pearl Harbor; at least forty-five other men from the Waterloo area joined the Navy in the weeks following the attack.[10] What made the Sullivan boys unique, however, was their insistence on being counted as one enlistment instead of five. When the recruiting officer could not promise the boys would be kept together, George Sullivan appealed to the Department of the Navy, requesting that he, his brothers, and two of their friends remain together for the duration of their assignment.[11] The unusual request required approval from multiple Navy officials and military personnel, but garnered the boys publicity and admiration in the interim. Promotional photographs, interviews, and posters were filled with images of the five Waterloo boys, their solidarity with one another and their resolve encouraging an anxious nation who was just coming to accept the reality of another world war.

On Saturday, February 14, the USS *Juneau* was officially commissioned. The cruiser was important for a number of reasons, but perhaps its biggest draw was that it was one of the first new ships to enter service following the attack on Pearl Harbor. The Navy needed to reassure the public following the attack, so an increased amount of pomp and circumstance surrounded their newest vessel.[12] Adding to the excitement of the new ship was the assignment of the "five husky Waterloo boys" who had captured the nation's attention a few months before.[13] The Sullivan brothers proved to be a publicist's dream for the Navy's recruiting campaign, never shying away from the camera or an eager reporter who wanted to hear more about the brothers. It was here that a press camera took one of the most well-known photographs of the Sullivan brothers. The five boys stood huddled together, grinning good-naturedly for the camera; there was no way of knowing this photograph would become one of the most tragic and widely distributed of the second World War.[14]

While press and Navy officials relished the publicity the boys received, not everyone was impressed with the Navy's decision to keep the brothers together. Lyman Swenson, the *Juneau*'s new captain, felt that the risk of family tragedy far outweighed the advantages of public relations.[15] He was proud of the Sullivan boys, and he was proud of the *Juneau,* but the possibilities for disaster were higher than he would like. The inherent danger of battle meant that one or several of the Sullivan boys could be injured or killed while in service, a fact which would inevitably hamper the media's fervor regarding the brothers. However, Swenson's concerns were casually waved away; the *Juneau* was a brand-new ship and she had a strong, if a little inexperienced, crew and there was no reason to think their assignment would be any different from the other cruisers that would eventually join it in the war.

This proved true for at least the first few months of the *Juneau*'s service. It engaged in typical wartime assignments, serving as an escort for supply ships and aiding in blockade efforts. The *Juneau's* first true voyage into the Pacific did not come until August, when a transmission requested the *Juneau* report to Vice Admiral Robert Lee Ghormley for an immediate assignment. Japan had maintained hegemony over much of Asia, including an arc of islands out into the Pacific Ocean. With a brutal convoy war taking place in the Atlantic Ocean, Japan sought supremacy in the Pacific by taking control of many of the islands and attempting to cut off open supply routes to the United States.

In May, the Japanese established their presence in the Solomon Islands and claimed Guadalcanal, the most strategically important property in the island chain. Allied forces landed on the island in early August and wrested control away from the Japanese. The victory was short-lived, however, as the Japanese retaliated almost immediately, sending waves of resistance against the Allied forces. Both sides dug in, each organizing strikes against the other's transport vessels and warships. The battle for Guadalcanal became one of the fiercest military campaigns in American history since the Civil War, with seven major naval engagements, at least ten land battles, and countless fire fights at sea and ashore.[16]

When the *Juneau* arrived in September, she was almost immediately thrust into the heat of battle. Just days before, the Japanese had torpedoed both the USS *Wasp* and the USS *Saratoga*. While the *Saratoga* was able to limp its way away from the battlefield, the *Wasp* was destroyed, taking over 200 men down with it. The *Juneau* filled in as its replacement, accompanying the USS *Hornet* northwest in an effort to stage an air raid against Japanese bases at Buin and Faisi in the Solomon Islands. During this time, the *Juneau* earned its first battle scars, taking minor damage during a number of air and sea battles. Its true crucible, however, did not take place until November, when it was assigned as part of an escort unit for reinforcements to Guadalcanal.

The first attack came from above—over thirty Japanese planes sweeping the escort group in a hail of bullets. The *Juneau*, which had been serving as a protective screen, launched into battle and managed to take down six torpedo bombers on its own.[17] During the melee, a torpedo fired by the Japanese destroyer *Amatsukaze* struck the *Juneau's* forward fire room on the port side, disabling central fire control and the powered gun turrets. Lester Zook, a sailor aboard the *Juneau*, recounted the experience, explaining, "this torpedo was very devastating . . . we lost all gunnery control and steering control . . . we were caught in the crossfire."[18] Unable to continue, the *Juneau* limped its way to safety, its crew desperately tending to the damages and the wounded.

On Friday, November 13, shortly after 11 a.m., the *Juneau* took a death blow. Ironically, the *Juneau* had not been the intended target; rather, the fatal torpedoes were intended for the USS *San Francisco*, another cruiser attached to the escort unit. The *San Francisco*, which had not been damaged as severely as the *Juneau*, was moving faster than its fellow cruiser, a fact which saved one ship, yet doomed the other. The torpedoes missed the *San Francisco* cleanly and instead found their mark in the crippled *Juneau*. The crew had seconds to realize what was about to occur, to try to react, but the actions were be fruitless.

The torpedoes struck in almost the same place the *Juneau* had been hit previously and a tremendous explosion ensued. Survivors and eyewitnesses would later recount that the *Juneau* did not sink, it disintegrated. In an official report from Lieutenant Commander Bruce McCandless, "the *Juneau* didn't sink. She blew up with all the fury of an erupting volcano."[19] Those below deck were killed instantly; the lucky few who were able to escape the blast found themselves tossed into the oil-slicked ocean. Acrid smoke plumed into the air and the *Juneau* sank in less than twenty seconds, with nearly 700 men still aboard.[20] Among those men were the Sullivan brothers.

News about the *Juneau's* fate did not reach the home front until near-

The Sullivans of Waterloo Iowa (left to right) Joseph (24), Francis (26) Albert (20), Madison (23), and George (28). Source: Naval History and Heritage Command, Catalog No. NH 52362.

ly two months later when the Sullivan family received a visit from Lieutenant Commander Truman Jones. Jones, who had sworn the boys into service a year earlier, now had the herculean task of informing the remaining Sullivan family that all five of their boys were missing in action and presumed dead. Aletta had heard some dark rumors around town regarding the *Juneau*, but Jones's visit confirmed her worst fears: her boys were gone, swallowed up by the cold, unforgiving sea. The news hit like a body blow for the Sullivan family and word spread quickly; within hours of Jones's visit the first reporters and news cameras arrived at the Sullivans' doorstep. Later that afternoon, an official Navy spokesman would address the nation, relaying information about the *Juneau* and the fate of its crew, including the Sullivan brothers. Shock and dismay led to outpourings of sympathy, and the Sullivan family suddenly found themselves at the uncomfortable center of a story they never wanted to be a part of.

What happened next would set the stage for how the nation coped with the tragedy and its perception of the continued war effort. The Sullivan family defined wartime loss and sacrifice in such personal terms that their reaction to the news became the framework through which the nation would view the war itself. The Sullivans, whether they wanted to or not, became some of the greatest influencers of the mid-war period simply based on their behavior. Had the family reacted negatively, refusing to meet with reporters and shutting out the waves of sympathy and government officials who were practically tripping over themselves to console them, the rest of the nation might have followed suit.[21] Public perception of the tragedy rested almost entirely on the Sullivan family's shoulders, and if the family felt that the loss was too much, that their boys had died for nothing, the nation would react in kind. A loss of support for the war effort back home would have crippled the Allies and ruined the chances of victory on the frontlines, and it was a risk the United States government and military simply could not take.

When the family did agree to meet with news reporters, the interviews were structured to promote triumph rather than tragedy. Rather than focusing on the deaths of the boys themselves, news reports emphasized the family's desire to persevere, to overcome their grief, and continue on with the good fight. As historian John Bodnar explains, the language used around and about the Sullivan family formed the parameters for how the rest of the nation should commemorate them; this family should not be pitied but lauded.[22] This message, which has been around since antiquity and has been used time and again to justify the killing of citizens for the national good, required the nation to accept the grim reality that casualties of war are necessary for victory. Had the Sullivan family entertained the notion that their sons' deaths were pointless and had little immediate effect on the war effort, they would never have been able to justify continued efforts back home. Instead, they were fed the carefully crafted narrative that the loss of their boys was tragic, yes, but that their sacrifice need not be in vain, and should instead contribute to continued war production on the home front. The phrase "My boys did not die in vain" became Aletta's battle cry for the rest of her life, a message she had to believe in order to continue on herself.

At the Navy's behest, Tom and Aletta soon embarked on a war production tour that would take them across the country with the intention to boost morale and encourage increased production on the home front. The tour also gave the media more time to shape the Sullivan family to better fit in the public's eye. Civilians on the home front experienced military conflict through carefully constructed signs and symbols, and their impressions were not so much derived from personal experience but from how they were rendered at home.[23] The Sullivan family effectively served as a gateway for the public to discuss the realities of war and their feelings about it. Specifically, it gave the nation a chance to analyze the relationship between combat and what was at stake back home.

The production tour succeeded in driving home the government-created narrative. The nation's focus quickly shifted away from the loss of the *Juneau* and was instead redirected to the legacy of Sullivan brothers and their solidarity as a family. According to Aletta, the boys would always end their letters with the simple phrase, "Keep your chin up,"[24] and it was a motto the nation latched onto through much of the remainder of the war. Every news story and interview focused on the bravery and courage of not only the Sullivan boys themselves but also their family. It was exactly the kind of message the nation needed during the mid-war period when battle fatigue and personal doubt were beginning to drag down national morale. When the war production tour ended in late 1943, a rousing success on all fronts, the Sullivan family retired back to Waterloo and attempted to return to life as usual, however impossible that may have been. There was one more chapter to the Sullivan story, however—one which catapulted the Iowa family onto the silver screen.

World War I launched the birth of a new genre of media: the war film. From the late 1920s through the early 1930s, Hollywood film emerged as the most popular form of entertainment in the United States and offered a release from the stresses of day-to-day life. Movies also served as one of the best forms of mass communication, effectively shaping morality, politics, and national attitudes toward social problems through the lens of film.[25] War films, in particular, offered a glimpse into an arena that was seldom viewed by regular citizens, but who were nonetheless desperate to understand war within a broader context. These films occupied a broad spectrum, includ-

Theatrical poster for the 1944 release of the 20th Century Fox feature film *The Sullivans*. Multiple versions of the film's poster were produced with the film's title interchangeably referred to as *The Fighting Sullivans* and *The Sullivans*. Poster produced in conjunction with 20th Century Fox, February 3, 1944.

Cover art for Dan Kurzman's 1995 book *Left to Die: The Tragedy of the USS Juneau.* The book features eye witness accounts from survivors and rescuers and provides firsthand accounts of *Juneau's* final battle and subsequent sinking. Kurzman, Dan. *Left to Die: The Tragedy of the USS Juneau.* New York: Pocket Books, 1995.

ing everything from newsreels and cartoons to instructional videos that were designed specifically for the home front. As American opinion was heavily shaped by the movies and films they viewed, it became abundantly clear that the stories and messages being presented needed to be carefully crafted.

In mid-1942, President Franklin Roosevelt developed a propaganda agency that would focus almost entirely on the film and media industry. The Office of War Information worked in conjunction with the Bureau of Motion Pictures to ensure that all films and media created during the mid-war period reflected the war in a positive light.[26] Public opinion had waned in the year since Pearl Harbor, and many were beginning to question the necessity of American involvement on the European front. As a result, the films that were released during this time were often unabashedly patriotic, their goal to return the American public to the war's favor. The three F's —faith, family, and freedom—were pushed in every film, along with the assurance that everyone must do their part to ensure the preservation of these ideals. Rather than light-hearted forms of entertainment, these persuasive war films were designed strategically for consumers and audiences and often had a very clear takeaway message. The Sullivan saga, with all its tragedy and triumph, served as the perfect vehicle to spread a wider message about the war itself.

Hollywood could not resist the Battle of Guadalcanal and within days of the announcement of the loss of the *Juneau* and the five Sullivan brothers,

Cover art for Issue #8, Vol. 10 edition of Male Magazine. The image depicts not only the sinking of the Juneau in the background but also the threat of sharks the survivors were left to contend with while awaiting rescue. Cover art created in conjunction with Male Magazine Corp., October 1958. "Shark Attack: Survivors of the USS Juneau." The Official Mort Künstler Website, October 1958.

Photo of the five Sullivan brothers at their home on Adams Street in Waterloo, IA. Photo is part of a collection belonging to the *Longview News-Journal*. The collection features photographs, news articles, and declassified Naval documents regarding the Sullivan brothers. "Remembering Waterloo's Five Sullivan Brothers." *Longview News-Journal*. November 12, 2021.

Photo of the film cast for the 1944 20th Century Fox film *The Fighting Sullivans*. The film depiction of the Sullivan family, while sincere, was vastly different from the real lives of the Sullivans and had been cleaned and polished to promote a pre-set image of the family. Hemenway-Forbes, Meta. "Sullivan Brothers' Story Showcased in Film, TV Productions." Waterloo Cedar Falls Courier, November 10, 2017.

Photo of *Juneau* survivor LCDR Lester Zook running the semaphore flags aboard the new USS The Sullivans during commissioning week in May 1997. Zook was one of ten men to survive the sinking of the USS Juneau, passing away at the age of 80 in 1998. "Remembering Waterloo's Five Sullivan Brothers." *Longview News Journal*. November 12, 2021.

moviemakers were already considering how they could put the catastrophe on the big screen.[27] The film, aptly named *The Fighting Sullivans*, premiered in February 1944, almost exactly a year after the beginning of the war production tour. It was intended to pay tribute to an "all-American family and their devotion and loyalty to themselves and their country."[28] In keeping with the wholesome American image the nation had devoured during the war production tour, the film focused on the early lives of the Iowa darlings, not the tragedy which befell them. The film reinforced and sentimentalized the image of the Sullivan family to the public—their lives defined by faith, hard work, and love for one another. Yet this wholesome representation, while great on film, acted as a prime example of the "good war" propaganda that had proliferated in the media during this period.

World War II had an unprecedented impact on American culture and the "good war" narrative was among one of the most popular tropes. These films valorized fighting, celebrated heroism, and generated a renewed faith in the government and the military.[29] Conventional treatment of combat in such films shrouded death on the battlefield with an aura of nobility, the unspoken understanding that sacrifice for a higher cause, in this case civilization itself, is always justified. A "good war" film was relentlessly patriotic, and depictions of actual violence were moderated or removed altogether. Military setbacks and combat casualties were referenced but not outright shown for the purposes of maintaining morale, after all, there was still a war to win and any showing of the true, deep horrors of war had the potential to repel the public rather than galvanize them. Such tactics were common during the mid-war period, and the general consensus in film and television was that death should be acknowledged but not shown on screen.

The Fighting Sullivans was no different and in fact, devotes little more than a few minutes of screen time to any actual combat. When the *Juneau* does inevitably sink at the end of the film, taking the titular characters down with it, it is off the screen and away from the public eye. In placing the death offstage, so to speak, the film succeeds in transforming the story's disheartening closure into a tale of virtue and courage—the Sullivan brothers become heroic martyrs who provide martial inspiration for the rest of the country.[30] It is a neat and clean way to wrap up the brother's tragic story—their deaths commemorated and honored on screen with the clear indication that their legacy will enjoy a much longer life than they did. This message, however, was little more than a clever marketing tactic on behalf of the Office of War Information, which had a vested interest in keeping just how ugly the true story of the *Juneau* was out of the spotlight.

In reality, the sinking of the USS *Juneau* was one of the worst maritime disasters of the Pacific War. The pretty, off-screen deaths depicted in *The Fighting Sullivans* smacked of bitter irony compared to the actual horrors of the real event. Contrary to film and media interpretation, the *Juneau* did not sink with all hands still onboard; roughly 130 men managed to make it off the doomed vessel before it plunged beneath the waves. The men who survived suffered horrific injuries—many left with broken bones, severed limbs, and devastating burns from the force of the blast.[31] Some men died within hours, some within days; the rest were left to suffer and wait, hoping and praying for rescue that seemed like it would never come. Fears of retaliatory attacks and misinformation delayed rescue operations for the wounded survivors for over a week and a half. By the time rescue attempts were made, only ten men were plucked from the water alive.

In order to properly address the Sullivan brothers' deaths in their titular film, a certain level of source manipulation needed to take place before it could be presented on screen. Even though the audience was fully aware of the fate of the eponymous characters, the climactic scene needed to be carefully constructed in order to best convey the overall message of the film. On film, the sinking of the *Juneau* is not instantaneous; rather it happens slow enough that the captain is able to order his men to abandon ship. Four of the brothers are above deck and defiantly ignore the orders to rush down below deck to the infirmary where George had been laid up in bed with an injury. The brothers rally around him, struggling to get him off the cot while they still have time, proclaiming, "We can't go swimming without you!"[32] There's a tremendous explosion in the background, a rattle below deck, and the scene fades to black with the silent, unspoken understanding that all five perished in the blast. It's a tragic yet sentimental scene which fully encapsulated the brothers' original message to the public: no matter what happened, they were all going to stay together. It would be lovely if the scene were true, but the reality was far bleaker.

When the *Juneau* was struck, at least three of the Sullivan brothers were killed instantly as the torpedo ripped through the body of the ship. Contrary to his cinematic counterpart, George Sullivan was actually above deck rather than below and was subsequently thrown into the water when the ship was torpedoed. Naval mechanic Allen Heyn recalled that he ended up in a life raft with George, the oldest Sullivan sibling still shell-shocked and covered in oil, pitifully searching the waters and calling out for his brothers.[33] Although Heyn remembered hearing a few flimsy reports that one of the other brothers survived, no trace of any of the other Sullivan siblings could be found. The other four Sullivan brothers had perished, and George was the sole survivor.

Following the sinking, the survivors of the *Juneau* faced days of relentless pain, thirst, and hunger. The sun broiled overhead and the men in the water, many of whom were still covered in oil from the explosion, were sunburned mercilessly.[34] Then, as if the ordeal had not already been traumatic enough, the survivors were then left to contend with another more terrifying threat. Heyn explained that by the third day, after most of the oil in the water had dissipated, the men began to no-

Collage photo of the Sullivan family including the five Sullivan brothers and their mother, Alleta. Aletta earned the unenviable title as a Five Gold Star Mother following the loss of all five of her sons aboard the *Juneau*. Gannon, Joe. "Alleta Sullivan: 'Champion Gold Star Mother' of WW2." The Wild Geese, November 22, 2018.

tice fins in the water. In the beginning, the sharks kept their distance, curiously circling the rafts full of men as they floated along the currents. When the oil cleared, however, and the smell of blood oozing from numerous open wounds grew stronger, the sharks moved in. Heyn made it clear that the smaller sharks weren't the issue; they swam closer to the surface and could be deflected with a well-placed kick or punch.[35] It was the larger sharks that proved to be the most terrifying menace. According to Lt. Commander Lester Zook, the larger sharks would strike from below, "chomping down on a leg or an arm . . . and dragging the [man] away."[36]

Once again, the film is careful to leave out the more gruesome aspects of the *Juneau* story, the on-screen depiction completely erasing what actually happened to the only surviving Sullivan brother. In a tragic, yet altogether anticlimactic end, Allen Heyn described the last moments of George Sullivan's life, claiming, "He said he was going to take a bath. And he took off all his clothes and got away from [the raft] and the white of his body must have flashed and showed up more because a shark came and grabbed him and that was the end of him. I never saw him again."[37] As these details would do little to promote a positive, sympathetic response from the American public, they were carefully scrubbed away and replaced with a more uplifting ending. In doing so, the off-screen deaths do more than blunt the tragedy of the *Juneau*, it essentially erased it entirely.[38]

The *Sullivan* film mentions nothing about the aftermath of the tragedy and the trauma it inflicted upon the survivors. In fact, it doesn't mention the survivors at all, a slight which infuriated Lester Zook.[39] "Friendly planes were flying over us daily," Zook recalled during an interview. "If help had arrived sooner there would have been a lot more of us alive."[40] In reality, a series of communication errors is what ultimately led to the delayed rescue effort. The force of the blast and the speed with which the *Juneau* sank led eyewitnesses on the nearby ship, the *San Francisco*, to believe that chances of survival were almost impossible. As such, early transmissions reported no survivors, and all nearby ships were ordered to pull back for fear of further attacks from the Japanese submarines still in the area.[41] Even when it became clear that there were men still alive in the water, it took several long days before rescue operations were approved. Knowing that news of military negligence and miscommunication would inevitably lead to public outcry, many of the awful details of the *Juneau* were largely covered up and downplayed.

Additionally, the film needed to present the surviving Sullivan family as the public saw them following news of the tragedy—as paragons of strength and courage in the face of unimaginable loss. Toward the end of the film, when the family is inevitably informed of the devastating news, their grief is quickly repackaged into heartrending dignity and stoicism, their suffering used as a model for thousands of other American families who had to face the cold reality of losing a loved one to war.[42] The family's plight still needed to be addressed, but in the proper manner; again, if the fictional Sullivans expressed any doubt in the necessity of the brothers' deaths, it could lead to a nationwide dip in morale and enthusiasm in the war effort. The film family's duty, then, was to transfigure the Sullivan family's sacrifice into a morale-boosting saga that would essentially ennoble combat casualties.

What the film could not address, however, was the full burden of grief real military families faced in the aftermath of tragedy. The process of grieving is highly culturally specific, and in the United States, there are certain actions that must be taken in order to fully accept the loss of a loved one.[43] In particular, viewing the body of the deceased helps bring home the reality and the finality of death to the family.[44] When someone is killed in such a way that body recovery is all but impossible, such as with the *Juneau*, accepting the reality of death is much harder for the surviving family. Naval conflicts are particularly difficult and in instances like the *Juneau*, the *Indianapolis*, and even the *Arizona*, where large numbers of fatalities occur all at once, a symbolic burial at sea takes place. However, while the deaths are acknowledged and commemorated, the families of those lost frequently find themselves in a kind of mental limbo, something author Michael Sledge refers to as "face in the crowd" syndrome.[45]

When a family is told that their loved one is dead, yet there is no body to bury, they will naturally harbor some small doubt in the veracity of the statement, constantly wondering if it was possibly a case of mistaken identity or assumption.[46] The common sentiment from families who experience this is that they cannot walk through a crowd without searching the faces for their missing or lost loved one. The lack of finality transforms into heartbreaking hope, something that can last a lifetime. The Sullivan family experienced this too, plagued with "what if" scenarios that lasted for years after the tragedy. Aletta herself admitted to this, explaining to one reporter, "I still have hope . . . maybe one of them survived. Maybe he is floating on a raft somewhere and he'll eventually make it back home."[47]

When citizens are killed in military service, the wounds are not localized and reserved only for the family; they extend to the body politic of the nation.[48] A nation's sense of self is partly defined by the strength of its military, so much so that service casualties create a sense of unbalance that begs for resolution. At a symbolic level, soldier death is representative of military service and, as such, embodies a specific kind of ideology, political belief, and culture.[49] Death has to mean something in the grand scheme of things and, particularly in the Sullivan brothers' case, it needs to feel necessary. The issue, then, as far as the *Sullivan* film was concerned, was not *if* the deaths should be covered, but *how*. The most intrinsic aspect of the "good war" narrative is that sacrifice and loss are never in vain, they are simply uncomfortable truths necessary to obtain a greater goal. Because of this, the Sullivan boys had to die a "good" death, one that could be praised and lauded rather than pitied.

To accomplish this, the film placed much higher emphasis on the boys' family and home life rather than the war. The Sullivan boys, after all, were supposed to represent what was right and good in America, their courage and eagerness to defend their country instantly transforming them into public sweethearts. The film needed to encapsulate this idea, it needed to reinforce the primacy of family and patriotism that was so intrinsically linked to the Sullivan's story. The film, therefore, would not serve as a war film so much as it would serve as a family biopic. Rather than focus on the war itself, it turned its attention to showing the boys enjoying a traditional, Midwestern upbringing complete with an idyllic childhood in small town America. It presented a rounded and deeply sentimental portrait of the five rough-and-tumble Sullivan brothers and their firm but loving parents. The characters were shaped and crafted in such a way that audience members of every age could relate to them; the boys could have been your classmates, their mother could have been your neighbor. But the focus on Waterloo and the mundanities of daily life were put on screen for a reason. The entire purpose of the film was to present what could not be ignored in mid-war America—the evils of the war could touch even the best of families.

Wartime is viewed as the ultimate disruption of normal family life and media, much like the kind centered on the Sullivan family, was designed to focus less on the traumatic impact of what had taken place and more on the family's traditional values of loyalty, faith, and patriotism.[50] Even in the earliest news stories and interviews, the Sullivan family was shaped and reframed to paint the family in the most glowing terms. War itself can only be understood, interpreted, and justified through the images and narratives connected to it, and the Sullivan family became paramount in conveying a care

Photo of the *USS The Sullivans*, a DD-537 destroyer launched in 1943 and christened by Alleta Sullivan in honor of her sons. The Sullivans was the first ship commissioned by the U.S. Navy that honored more than one person. The ship's badge includes the Sullivan brother's oft-repeated motto, 'We Stick Together.' Powell, Eliza. "Sullivan Family Reunites on USS The Sullivans in Mayport to Honor 5 Sullivan Brothers Killed during WWII." September 27, 2019.

fully constructed message to the public. Similarly, the Sullivan family on screen had to embody that same public image, one that affirmed God, country, and family, and the social values that rationalized the war effort both back home and on the frontlines; in short, the Sullivan family had to represent the bedrock of American culture.[51]

While the film itself proved to be remarkably popular with American audiences, elements of Hollywood myth-making could not be ignored. All too frequently, "good war" films fell into the dangerous terrain of mythical sanitization, the conventions of melodrama making for a better story while obscuring some of the more brackish realities. In order for the nation to fully accept the Sullivans and the message they were spreading, their image needed to be adjusted. The less savory aspects of the family's lives were painted over and were instead primed and polished to make the Sullivans appear as the perfect Midwestern family. The fact that the Sullivan family had been cleaned up for the public's eye was not the issue; the problem was that the family onscreen barely resembled the family in real life. Even surviving members of the family were surprised by how heavily their lives had been sanitized for the sake of the film. Katherine McFarland, Albert Sullivan's widow, remarked that the film "was just a lot of BS. Everyday life was too boring, they had to spice it up a little bit."[52]

The film family was so squeaky clean, in fact, that at times they came across as maudlin and cloying.[53] The film never made mention of Tom's well-known alcohol abuse or Aletta's spiraling bouts of depression prior to the war, and it certainly never addressed the fact that all five Sullivan boys were high school dropouts who regularly got into fistfights and would purposely wander into the African-American neighborhoods in Waterloo for the express purpose of starting racially-fueled fights.[54] The public also didn't need to know that the youngest brother, Albert, potentially had a child out of wedlock and then enlisted in the Navy along with the rest of his brothers because he was "looking for a way out."[55] Anything that could have diminished the Sullivan family's image to the public was carefully screened out in favor of a more wholesome, traditional depiction on screen.

While the discrepancies between the real Sullivans and their silver screen counterparts seem questionable now, the sanitized version was a necessary marketing ploy in order to keep the stakes of the war meaningful. The nation needed to believe in something and for a while the Sullivan family was it. The intersection of myth and memory was frequently blurred in *The Fighting Sullivans*, but it was necessary to give the public a reason to keep fighting. Because a fundamental aspect of war involves destruction and death, allowances needed to be made within the cultural sphere of what was acceptable within that framework. Death without reason was unacceptable, but death for a cause wasn't. Both the real and the fictional Sullivan boys answered a divine call, standing up in the face of unimaginable danger to protect all the things they held near and dear. Their cause was not only each other but also their freedom and family back home. It was exactly the kind of message the government wanted to convey, and the *Fighting Sullivans* was the perfect film to do this. This, ultimately, is what would transform a simple Iowan family into symbolic heroes.

What is interesting about the Sullivan saga is how extensively it defined the parameters for wartime heroism. The world needed heroes during the mid-war period more than ever, but in order to gain them, ordinary people had to be elevated beyond the normal scope of life. It would be enough to confer heroism among individuals because of their bravery and merit, but the demand for perfectionism was an all too frequent hurdle in the mid-1940s. The American public latched onto the Sullivan brothers because they represented an ideal, but even that wasn't enough; their image had to be reshaped and cleaned up to fit a pre-packaged narrative. The Sullivan family could not hope to be remembered unless their social origins and public presentation were deemed worthwhile by the American public. Sympathetic audiences needed to see their trials and tragedy as undeserved. If the public did not respect and embrace the Sullivan family, why should they care about the boys' deaths?

The American public recognizes the need for moments of silence only when bad things happen to good, upstanding people. Hollywood and the United States government, who had made paragons of the Sullivan family, all but demanded the public acknowledge their sacrifice because they were good people who were victims of bad things. The mid-war period was fraught with anxiety and doubt and because of that, the people perceived as heroic needed to be pristine. Truth was frequently overwritten with fiction and heroic narratives only needed to be presented as worthwhile, not accurate. It is because of this essential need for heroes that the Sullivan story needed to be adjusted for the public eye. The story of the Sullivan brothers is extraordinary, that much is clear, but it was Hollywood myth-making that transformed them into heroes.

What Hollywood saw in the Sullivan family was an opportunity, a way to speak to the public on a more personal level. The Sullivan boys had captured the nation's attention and, following their untimely deaths, their surviving family had captured their hearts. The Sullivan saga was unlike anything the world had seen before and represented a new and effective way of keeping the public invested in the war effort. Although their story bore little resemblance with the story that would eventually make its way to the big screen, the five brothers served as the perfect backdrop and represented the prop in "good war" propaganda. When their ship was finally located off the Solomon Islands on March 17, 2018, it seemed the nation could finally rest in the knowledge that the boys' final resting place had been found.[56] With all wars, fragments of myth and memory are conjoined to render meaning to the fighting and bloodshed. Hollywood mythmaking transformed the Sullivan brothers, nearly deifying them in the realm of cinema, but it was necessary to produce an accepted cultural understanding of the mid-war years.

Works Cited

Anderson, Jack, and Michael Binstein. "A 52-year-old Coverup of Navy Negligence." *The Washington Post*. May 30, 1994. https://www.washingtonpost.com/archive/lifestyle/1994/05/30/a-52-year-old-coverup-of-navy-negligence/3467e4d8-2b64-4d42-8af8-b1279dbe77fb/?utm_term=.241c4c67bf0a.

Andersen, Robin. *A Century of Media, a Century of War*. New York: Peter Lang, 2007.

Bodnar, John E. *The "Good War" in American Memory*. Baltimore, MD: Johns Hopkins University Press, 2012.

Daley, Jason. "Wreck of U.S.S. Juneau Discovered in the Solomon Islands." Smithsonian.com. March 22, 2018. https://www.smithsonianmag.com/smart-news/wreck-uss-juneau-discovered-solomon-islands-180968556/.

Heyn, Allen. "One Who Survived."" Interview. *American Heritage*, June 1956. http://www.americanheritage.com/one-who-survived?page=show.

"Juneau I (CL-52)." Naval History and Heritage Command. https://www.history.navy.mil/research/histories/ship-histories/danfs/j/juneau-i.html

Kinney, Pat. "Brothers in Arms: Remembering the Five Sullivan Brothers, 75 Years Later." *Waterloo Cedar Falls Courier*. November 10, 2017.

Kuklick, Bruce. "A Band of Brothers: The Story of the Sullivans." HistoryNet. May 10, 2017. https://www.historynet.com/a-band-of-brothers-the-sullivans.htm.

Kuklick, Bruce. *The Fighting Sullivans: How Hollywood and the Military Make Heroes*. Lawrence, KS: University Press of Kansas, 2016

Kurzman, Dan. *Left to Die: The Tragedy of The USS Juneau*. New York: Pocket Books, 1994.

McDowell, John. "The Terrible Ordeal of the USS Juneau." *Sea Classics* 19, no. 2. March/April 1986.

McFarland, Katherine Sullivan. "Albert Sullivan's Widow Looks Back." Interview by Pat Kinney. *The Courier*. November 11, 2012. https://wcfcourier.com/news/local/albert-sullivan-s-widow-looks-back/article_e0e04cf3-954d-51d8-817d-3d8f-da4d67f8.html.

Morison, Samuel Eliot. *History of United States Naval Operations in World War II*. Vol. 5. *The Struggle for Guadalcanal*. Little, Brown and Company, 1950.

Niemi, Robert. *History in the Media: Film and Television*. Santa Barbara: ABC-CLIO, 2006.

Satterfield, John R. *We Band of Brothers: The Sullivans and World War II*. Parkersburg, IA: Mid-Prairie Books, 2000.

Sledge, Michael. *Soldier Dead: How We Recover, Identify, Bury, and Honor Our Military Fallen*. New York: Columbia University Press, 2007.

"Sullivan Mother Carries On." *The Lowell Sun*, January 21, 1943.

"Sunken USS Juneau Famous for the Sullivan Brothers Discovered on St. Patrick's Day." Paul Allen. https://www.paulallen.com/uss-juneau-wreck-located-famous-for-sullivan-brothers/.

Talbott, Basil. "Fighting Sullivans: Brave Mother Carries On." *Chicago Herald-American*, January 20, 1943.

The Fighting Sullivans. Directed by Lloyd Bacon. United States: Twentieth Century Fox, 1944.

Worden, J. William. *Grief Counseling and Grief Therapy: A Handbook for the Mental Health Practitioner*. New York, NY: Springer Publishing Company, LLC, 1991.

Notes

1 "Sunken USS Juneau Famous for the Sullivan Brothers Discovered on St. Patrick's Day." Paul Allen. March 17, 2018.

2 Talbott, Basil. "Fighting Sullivans: Brave Mother Carries On." *Chicago Herald-American*, January 20, 1943.

3 Satterfield, John R. *We Band of Brothers: The Sullivans and World War II*. (Parkersburg, IA: Mid-Prairie Books, 2000), 54.

4 Ibid.

5 Kuklick, Bruce. *The Fighting Sullivans: How Hollywood and the Military Make Heroes*. (Lawrence, KS: University Press of Kansas, 2016), 29.

6 Kinney, Pat. "Brothers in Arms: Remembering the Five Sullivan Brothers, 75 Years Later." Waterloo Cedar Falls Courier. November 10, 2017.

7 Kuklick, 36.

8 Satterfield, 55.

9 Kuklick, 38.

10 Satterfield, 55.

11 Kuklick, 39.

12 McDowell, John. "The Terrible Ordeal of the USS Juneau." *Sea Classics* 19, no. 2 (March/April 1986), 12.

13 Kuklick, 40.

14 Satterfield, 67.

15 Ibid.

16 Morison, Samuel Eliot. *History of United States Naval Operations in World War II*. Vol. 5. The Struggle for Guadalcanal. (Little, Brown and Company, 1950), 164.

17 Satterfield, 130.

18 Zook, Lester; quoted in Satterfield, *We Band of Brothers*, 135.

19 McCandless, Bruce; quoted in Satterfield, *We Band of Brothers*, 148.

20 "Juneau I (CL-52)." Naval History and Heritage Command.

21 Satterfield, 169.

22 Bodnar, John E. *The "Good War" in American Memory*. (Baltimore, MD: Johns Hopkins University Press, 2012), 110.

23 Andersen, Robin. *A Century of Media, a Century of War*. (New York: Peter Lang, 2007), xvi.

24 Satterfield, 171.

25 Kuklick, 96.

26 Ibid, 101.

27 Ibid, 106.

28 Kuklick, Bruce. "A Band of Brothers: The Story of the Sullivans." HistoryNet. (May 10, 2017), 5.

29 Anderson, 27.

30 Niemi, 76.

31 Kurzman, Dan. *Left to Die: The Tragedy of The USS Juneau*. (New York: Pocket Books, 1994), 165.

32 *The Fighting Sullivans*. Directed by Lloyd Bacon. (United States: Twentieth Century Fox, 1944), 1:44:04.

33 Kurzman, 166.

34 Heyn, Allen. "One Who Survived." Interview. (*American Heritage*, June 1956), 8.

35 Heyn, 7.

36 Zook, Lester; quoted in Satterfield, *We Band of Brothers*. 155.

37 Heyn, 10.

38 Bodnar, 110.

39 Kuklick, 73.

40 Anderson, Jack, and Michael Binstein. "A 52 year old Coverup of Navy Negligence" *The Washington Post*. May 30, 1994.

41 "Juneau I (CL-52)." Naval History and Heritage Command.

42 Ibid.

43 Sledge, Michael. *Soldier Dead: How We Recover, Identify, Bury, and Honor Our Military Fallen.* New York: Columbia University Press, 2007.

44 Worden, J. William. *Grief Counseling and Grief Therapy: A Handbook for the Mental Health Practitioner.* (New York, NY: Springer Publishing Company, LLC, 1991), 61.

45 Sledge, 269.

46 Ibid.

47 "Sullivan Mother Carries On." *The Lowell Sun*, January 21, 1943.

48 Sledge, 292.

49 Ibid, 26.

50 Bodnar, 110.

51 Niemi, Robert. *History in the Media: Film and Television.* (Santa Barbara: ABC-CLIO, 2006), 76.

52 McFarland, Katherine Sullivan. "Albert Sullivan's Widow Looks Back." Interview by Pat Kinney. *The Courier*. November 11, 2012.

53 Niemi, 76.

54 Kuklick, *A Band of Brothers*, 3.

55 Kuklick, Fighting Sullivans, 36.

56 Daley, Jason. "Wreck of U.S.S. Juneau Discovered in the Solomon Islands." Smithsonian.com. March 22, 2018.

Confederate Mission Command at Chickamauga: A Case Study of Braxton Bragg and A.P. Stewart

Matthew A. Hughes, Captain, U.S. Army
American Military University

The opinions expressed in this article are those of the author alone and do not reflect the official position of the U.S. Army, the Department of Defense, or the U.S. Government. Also, the appearance of hyperlinks does not constitute an endorsement by the U.S. Army, the DoD, or the U.S. Government of the referenced sites or the information, products, or services contained therein.

Abstract

In the American Civil War, the Battle of Chickamauga of 18–20 September 1863, holds important lessons on mission command, particularly the actions of Confederate General Braxton Bragg and one of his division commanders, A.P. Stewart. Their degree of application of the principles of mission command greatly influenced the course of the battle. Bragg's incompetency in utilizing favorable terrain, discord in relationships with subordinate commanders, and vague or shifting mission orders caused the Confederacy to squander opportunities to destroy the Union army in early September and at other points that could have proved decisive during the battle. In contrast, Stewart's tactical competency and mutual trust among his subordinates enabled his division to seize key terrain and penetrate Union defenses that adjacent units could have exploited for an earlier and more decisive victory.

Keywords: Chickamauga, Chattanooga, Alexander P. Stewart, Braxton Bragg, American Civil War, Mission Command, Confederate States Army, Leadership

Mando de la misión confederado en Chickamauga: un estudio de caso de Braxton Bragg y A.P. Stewart

Resumen

En la Guerra Civil Estadounidense, la Batalla de Chickamauga del 18 al 20 de septiembre de 1863 contiene lecciones importantes so

doi: 10.18278/sshj.10.3.6

bre el mando de la misión, en particular las acciones del general confederado Braxton Bragg y uno de sus comandantes de división, A.P. Stewart. Su grado de aplicación de los principios del mando de la misión influyó mucho en el curso de la batalla. La incompetencia de Bragg para utilizar un terreno favorable, la discordia en las relaciones con los comandantes subordinados y las órdenes de misión vagas o cambiantes hicieron que la Confederación desperdiciara oportunidades de destruir el ejército de la Unión a principios de septiembre y en otros puntos que podrían haber resultado decisivos durante la batalla. En contraste, la competencia táctica de Stewart y la confianza mutua entre sus subordinados permitieron que su división tomara terreno clave y penetrara las defensas de la Unión que las unidades adyacentes podrían haber explotado para una victoria más temprana y decisiva.

Palabras clave: Chickamauga, Chattanooga, Alexander P. Stewart, Braxton Bragg, Guerra civil estadounidense, Mando de la misión, Ejército de los Estados Confederados, Liderazgo

联盟军在奇卡莫加战役中的任务式指挥：关于布拉克斯顿•布雷格和亚历山大•P•斯图尔特的案例研究

摘要

美国内战期间，1863年9月18-20日发生的奇卡莫加战役对任务式指挥具有重要意义，尤其是联盟军上将布拉克斯顿•布雷格及其下属师长之一的亚历山大•P•斯图尔特所采取的行动。他们对任务式指挥原则的应用程度极大影响了战役进程。布雷格在利用有利地形一事上的能力不足、与下属司令官关系的不和谐、以及模糊或变化的任务指令，导致联盟国浪费了在9月早期以及其他时刻击败合众国军队的机会，而这些时刻本有可能在战役中发挥决定性的作用。相反，斯图尔特的策略能力和与下属之间的相互信任使其军队抓住关键地形并击破合众国军队的防御，而其邻近部队本有可能利用有利地形争取更早、更具决定性的胜利。

关键词：奇卡莫加战役，查塔努加，亚历山大•P•斯图尔特，布拉克斯顿•布雷格，美国内战，任务式指挥，联盟军，领导力

The Influence of Mission Command

Unlike the statues and monuments at Gettysburg, Vicksburg, or Antietam that commemorate generals, those at Chickamauga honor lower-echelon leaders who provided direction when the top commanders of each side failed to exercise effective mission command during the battle fought between 18 and 20 September 1863.[1] One notable case of Confederate officers at the battle involved General Braxton Bragg, overall Confederate commander in the battle over the Army of Tennessee, and one of Bragg's division commanders, Major General Alexander Peter (A.P.) Stewart. In 1894, W. J. McMurray, a member of the 20th Tennessee Infantry Regiment—one of the five regiments in Stewart's division—called for "a monument to Stewart's memory" at the site where Stewart's "'Little Giant Division' broke the Federal center."[2] Later, at the Tennessee monument dedication on Horseshoe Ridge in 1895, James D. Porter, former Governor of Tennessee and Confederate veteran of this battle, declared that the Battle of Chickamauga would not have been a barren victory if Stewart had commanded the army rather than Bragg.[3] While this remark can be "dismissed as complimentary hyperbole,"[4] the statement stems from important contrasts between these two Confederate leaders and their different mission command capacities.

This article investigates the leadership styles and abilities of these two leaders at Chickamauga through the lens of contemporary Army doctrine on mission command, or "the Army's approach to command and control that empowers subordinate decision making and decentralized execution appropriate to the situation."[5] It seeks to answer the question, how did A.P. Stewart's leadership differ from Braxton Bragg's at Chickamauga? Primary source material including diary accounts from soldiers, official reports on the battle by leaders, and circumstantial evidence from how the battle progressed offer details on how these two fared in the principles of mission command: competence, mutual trust, shared understanding, commander's intent, mission orders, disciplined initiative, and risk acceptance (see Table 1). Ultimately, Bragg's weak mission command in terms of incompetence in utilizing terrain, poor mission orders, and lack of mutual trust hindered Confederate forces and contributed to missed opportunities to defeat the Union army; whereas Stewart's tactical competence, disciplined initiative, risk acceptance, and strong mutual trust enabled Confederate forces to secure key terrain and created opportunities to defeat the Union army.

After driving General Braxton Bragg's forces out of middle Tennessee in the summer of 1863, General William S. Rosecrans led his 60,000-strong Army of the Cumberland southward toward Georgia. His objective was to seize Chattanooga—a vital Confederate railroad hub that connected four major railroads and offered a route through the Appalachian Mountains deeper into Georgia. President Abraham Lin-

Principle	Definition/Description
Competence	Commanders, subordinates, and teams possessing tactical and technical abilities to perform their necessary tasks and functions.
Mutual trust	Shared confidence between commanders, subordinates, and partners that they can be relied on and are competent in performing their assigned tasks.
Shared understanding	A mutual comprehension of an operational environment, an operation's purpose, problems, and approaches to solving problems.
Commander's intent	A clear and concise expression of the purpose of the operation and the desired military end state that supports mission command, provides focus to the staff, and helps subordinate and supporting commanders act to achieve the commander's desired results without further orders, even when the operation does not unfold as planned.
Mission orders	The process of issuing directives to subordinates conveying desired results without specifying how to achieve those results.
Disciplined initiative	The duty individual subordinates have to exercise [enterprise or leadership] within the constraints of the commander's intent to achieve the desired end state.
Risk acceptance	Application of judgment to make choices with the potential for loss in return for certain benefits that would be unattainable otherwise.

Figure 1. Principles of Mission Command. Source: Generated by the author based on definitions and explanations in Department of the Army, Mission Command: Command and Control of Army Forces, ADP 6-0 (Washington, DC: Department, of the Army, 2019), https://armypubs.army.mil/epubs/DR_pubs/DR_a/ARN18314-ADP_6-0-000-WEB-3.pdf, 1-6 to 1-14.

coln considered this site to be strategic key terrain and just as vital as Richmond. Although Bragg had been outmaneuvered in Tennessee, his Army of Tennessee received reinforcements and totaled around 65,000 by the time Rosecrans approached Chattanooga. Bragg fortified Chattanooga and prepared for a Union attack from the north—not a maneuver to his west.

Within Bragg's army, A.P. Stewart led a division that consisted of the following components: Foule's Co. Mississippi Cavalry (escort); Johnson's Brigade (detached) with attached artillery (York's Georgia Battery); Bate's Brigade with attached artillery (Eufaula Alabama Battery); Brown's Brigade with attached artillery (Dawson's Georgia Battery); and Clayton's Brigade with attached artillery (1st Arkansas Battery).[6]

Precursory Events (15 August–17 September 1863)

Actions taken by Bragg and Stewart in late August and early September 1863 reveal important distinctions regarding their levels of competence, or the trait of "commanders, subordinates, and teams possessing tactical and technical abilities to perform their necessary tasks and functions."[7] Bragg carried with him a bleak string of defeats throughout the war and the Union's recent Tullahoma campaign in Tennessee. Bragg's Army of the Tennessee also suffered from a lack of cohesion and mutual trust, largely due to the lack of confidence for Bragg shared by his subordinate commanders and lowest-ranking troops alike.[8] Bragg's decisions in early and mid-September 1863 compounded this rancor in the ranks and distain of him for his perceived incompetence. Meanwhile, Stewart's preparations, tactical competence, and mutual trust formed a cohesive unit that would obey his command even in the direst circumstances.

Bragg's Incompetence regarding the Terrain and Employment of Forces

Bragg's inability to visualize the battlefield and utilize his resources effectively led him to lose several advantages the terrain afforded to his defense and enabled Rosecrans to pass through Lookout Valley to the west of Bragg's forces at Chattanooga undetected (see Figure 2).[9] Anticipating a Union attack from the north, he positioned reconnaissance forces north and along the Tennessee River, but neglected the heavily-wooded, rolling terrain westward. He also failed to capitalize on the Tennessee River as a natural obstacle to approaching Union forces, where he could have planned for indirect fires to disrupt Union forces as they approached and tried to cross the river.

Instead, Rosecrans marched through this terrain uncontested. His forces then crossed the Tennessee River southwest of Chattanooga and conducted a turning movement to threaten Confederate lines of commu-

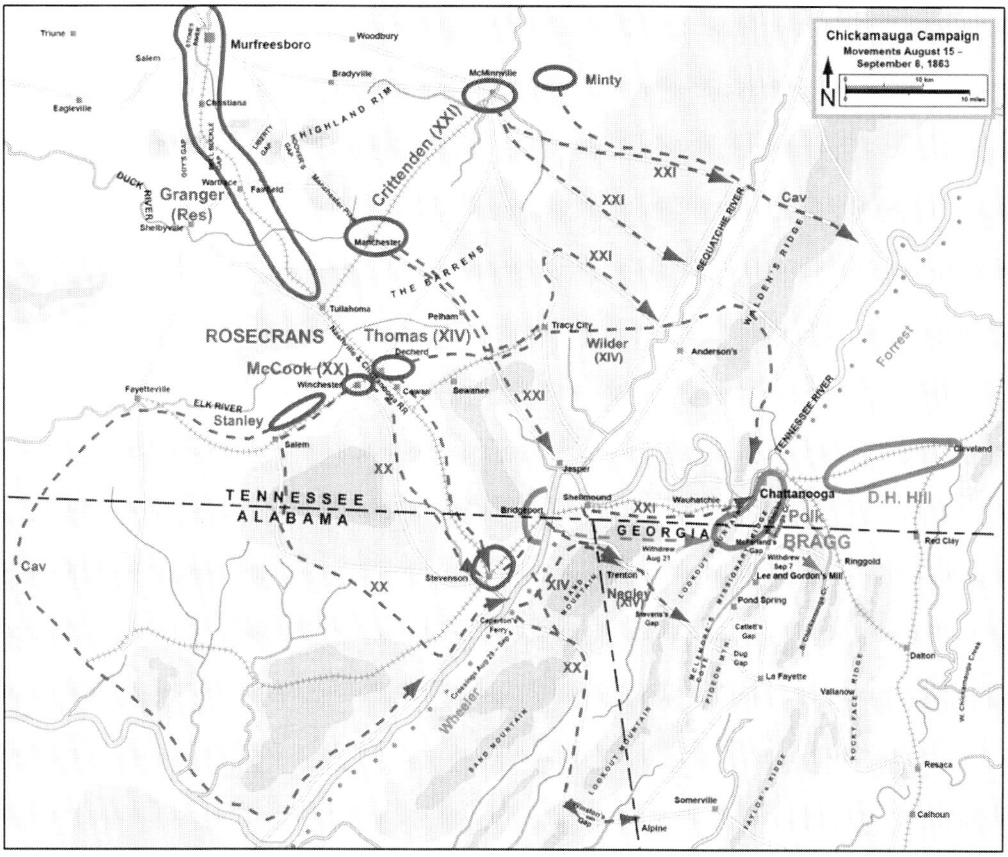

Figure 2. Initial Movements in the Chickamauga Campaign, 15 August - 8 September 1863. Source: Hal Jespersen, "Map of Chickamauga Campaign of the American Civil War," 2 November 2008, http://www.posix.com/CWmaps/.

nication, so as to draw Bragg's Army of Tennessee out of its defenses in Chattanooga. When the surprised Bragg was alerted to these Union forces to his south on 8 September, he did as Rosecrans intended and abandoned Chattanooga, despite opposition from D.H. Hill, one of his corps commanders.[10] He moved his army south to interdict Rosecrans and prevent the disruption of his lines of communication. Despite Bragg's reconnaissance failures and cession of Chattanooga (which the Union occupied on 9 September), the dispersed Union forces crossing the river presented Bragg with an opportunity to defeat them while they were vulnerable.

Squandered Opportunities due to a Lack of Mutual Trust under Bragg

The Union army split into three distinct columns of corps-sized elements as it advanced, utilizing the three routes suitable for movement. Bragg, aware of this dispersion across 65 km, planned to attack the separated corps one by one while they were vulnerable and too far from one other for mutual support.[11] At 9:00 p.m. on September 10, Bragg uncharacteristically convened a council of war with his division and brigade commanders to outline plans to attack Rosecrans's separated columns. The official record on this council remains unclear, but three possibilities exist regarding the meeting's outcome: (1) Stewart was unaware of Bragg's positive orders; (2) Bragg's subordinate commanders were aware of his orders and most understood them to not be discretionary, but Major General Thomas C. Hindman (senior Confederate officer of troops closest to Union forces under Negley) believed he had discretion on when to attack; or (3) that the generals felt they had discretion but that Bragg's orders were no longer worth obeying.[12]

Over the following days, the Confederates failed to attack these vulnerable Union corps in what is traditionally attributed to insubordination of Bragg's division commanders, a manifestation of the lack of mutual trust between Bragg and his subordinate commanders. Stewart's division participated in the potentially decisive isolation and defeat of one of Rosecrans's corps at Dug Gap, but Bragg failed in this attempt mainly due to his poor command relationship with subordinate commanders.[13]

Several of these subordinates later criticized Bragg for this failure. D.H. Hill blamed Bragg and his methods of command for failures to exploit Union weaknesses prior to the Battle of Chickamauga.[14] A.P. Stewart wrote the following in 1886 while reflecting on these missed opportunities in early September 1863: "Bragg was in a position to crush the enemy's center and interpose his army between the wings, which could not have escaped . . . Whatever apologies may have been offered for this failure, the real cause of it was the lack of confidence on the part of the superior officers of the Army of Tennessee in its commander. If Robert E. Lee or either of the Johnstons had been in command, the blow would have been struck, and in all human probability Rosecrans's army would have been destroyed."[15] Although Stewart rarely criticized Bragg or others openly, his writings clearly reveal his contempt for Bragg and blame his weak mission command abilities for these missed opportunities.

Competence and Mutual Trust Yield a Ready Force under Stewart's Direction

In contrast to the doubt held and disdain harbored by division commanders for Bragg, Stewart's peers and subordinates recognized Stewart's competence and leadership abilities. Philip Stephenson explained that Stewart was "never regarded by the men as having the qualities of greatness," but he rose steadily through the ranks due to merit—"he never seemed to make a mistake!"[16] Despite his station as the newest division commander at Chickamauga, he had the trust and confidence of superiors, peers and subordinates. B.L. Ridley, who had served as one of Stewart's staff officers, paid Stewart tribute when he said in 1895, "When other commanders found that Stewart was supporting them, on right or left," he said, "all was well; and when he struck the enemy, there were frequently heartrending scenes of carnage and of blood."[17] Ridley also commented that those who served under Stewart's command know that "he would not willingly sacrifice them, [and] whenever he said to do so, they would leap into the very jaws of death."[18] In the weeks prior to the Battle of Chickamauga, Stewart had taken measures to reestablish discipline in his unit through drill and issuance of orders on establishing and policing camps.[19] While at Chattanooga, it seems Stewart shared Bragg's belief that the Union army would cross the Tennessee River just north of Chattanooga, and so Stewart distributed his Brigades at crossing points in anticipation, nested with Bragg's defensive plans.[20]

Opening Engagements (18 September 1863)

In alignment with his envisioned envelopment of the Union army, Bragg ordered his army to extend north to where he thought the Union's left flank would be—at Lee and Gordon's Mill—and then to cut across West Chickamauga Creek to isolate the federals and prevent their retreat to Chattanooga. Bragg specified four crossing points for his elements: (1) Reed's Bridge for Johnson's division, (2) Alexander's Bridge for Walker's Reserve Corps, (3) Thedford's Ford for Buckner's corps (which included A.P. Stewart's division), and (4) Dalton's Ford for Polk's corps (see Figure 3).[21] Confederate units executed the orders and began to make contact with Union forces the morning of 18 September at unexpected points on the battlefield.

Persistent Efforts to Fulfill Bragg's Intent

The array of Confederate forces and locations of initial engagements on the 18th demonstrate that Bragg successfully conveyed his intent for his army to interdict Union movement northward to Chattanooga. Army doctrine describes commander's intent as "a clear and concise expression of the purpose of the operation and the desired mili

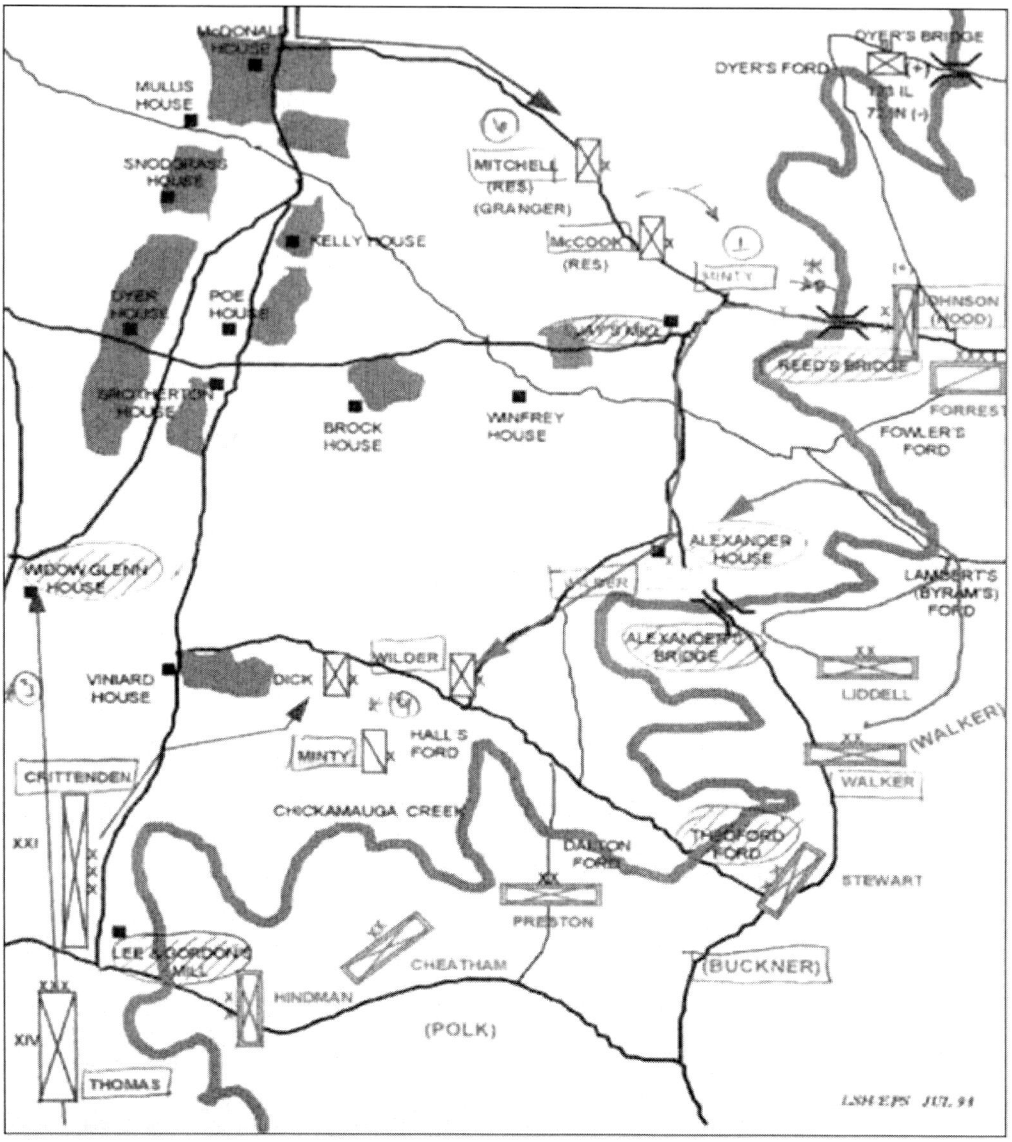

Figure 3. Movement to Contact, 18-19 September 1863. Source: "Estudio de Campo: La Batalla de Chickamauga (Case Study: The Battle of Chickamauga)," presentation at Command and General Staff Officers' Course (CGSOC), Western Hemisphere Institute for Security Cooperation (WHINSEC), Fort Benning, GA, 14 September 2021, Slide 10.

tary end state that supports mission command, provides focus to the staff, and helps subordinate and supporting commanders act to achieve the commander's desired results without further orders, even when the operation does not unfold as planned."[22] As Bragg's forces made initial contact around 7:00 a.m. near Reed's Bridge (see Figure 3)—further north than he expected Union forces to be—his right wing extended in an attempt to turn the Union and cut off its retreat toward Chattanooga.[23] Additional engagements to the north should have led Bragg to realize his assessment of the Union disposition was wrong and that he should reconsider the intent of turning the Union army, as this proved to be unfeasible. Instead, the Confederate right wing was reinforced by forces from the left wing and continued to extend northward. This led to the battle unfolding at a location and in a manner neither side had anticipated.

Stewart's Competence at Thedford's Ford

Stewart displayed tactical and technical competence at Thedford's Ford as he directed his division in a successful obstacle crossing while under fire. After receiving the order from his corps commander, Major General Simon B. Buckner, to occupy the key terrain dominating the ford but to avoid engagement with the enemy unless necessary,[24] Stewart arrayed his forces for the task. He tasked his engineer attachment to reconnoiter the site, led with his most experienced brigade, and maintained a reserve.[25]

Stewart also emplaced an artillery battery on the high ground to provide overwatch, which, along with the 4th Georgia Battalion of sharpshooters of Bate's brigade, later engaged Union skirmishers across the creek in what the artillerymen later claimed to be the opening shots of the Battle of Chickamauga.[26] The efficiency with which Stewart's division executed this task attests to Stewart's competence and the shared understanding of his subordinate commanders and staff, especially the discipline displayed by inexperienced soldiers under direct and indirect fire that ranged much of the area.[27] In fact, while Brigadier General John Pegram, commanding cavalry, and Stewart were discussing plans nearby, "a cannonball struck about five paces from them before careening onward and nearly killing a member of [Brigade Commander J.C.] Brown's staff."[28]

First Day of Battle (19 September 1863)

On the morning of the 19th, Union forces sought to secure a line of retreat northward while the Confederate line extended further northward in an attempt to locate the Union's left flank and turn the Union army. From the point of initial skirmishes, Rosecrans and Bragg slowly committed more forces until the Union formed a defensive line along Lafayette Road to repel Confederate attacks. Smoke from gunfire and cannon fire pervaded this line as the battle carried on into the afternoon. Bragg's poor mission command contributed to confusion among Confederate leaders and a lack of forward progress. Stewart, on the other hand, displayed mission command abilities that nearly split the Union defense into two separate forces and offered an opportunity for Bragg's army to defeat Rosecrans.

Bragg's Mission Orders

Bragg's mission orders on the 19th were limited due to his distance from the battlefield and vagueness stemming from reorganization of his forces and ad hoc decisions. First, Bragg established his command post in vicinity of Thedford's Ford, near the intersections of the main roads (see Figure 4). This considerable distance from his main forces, however, limited his ability to monitor progress of the battle and provide direction. In contrast, Rosecrans established his command post within a few hundred yards of the Union defensive line along Lafayette Road and was able to inspect his defensive line and convey orders throughout the 19th (although illness led him to delegate some of those authorities and limit his personal involvement in mission orders, which negatively impacted Union mission command; see Figure 4).[29] With such factors in the battlefield as heavily-wooded terrain, greater proximity to his subordinate commanders and units would have facilitated Bragg's mission orders and reevaluation of the operational picture throughout the battle.

Second, Bragg's ad hoc decisions and guidance, sometimes bypassing subordinate commanders in filtering down to divisions, led to confusion regarding mission orders. As Army doctrine states, a mission order is "a communication—verbal, written, or signaled—that conveys instructions from superiors to subordinates" and allows "maximum freedom of action in accomplishing missions."[30] Doctrine holds that the level of detail in mission orders will vary depending on the situation, but "they are neither so detailed that they stifle initiative nor so general that they provide insufficient direction."[31] As skirmishes erupted on the morning of the 19th, Bragg showed "little regard for the proper chain of command" as he issued orders and slowly committed forces to support those in contact, rather than amassing forces.[32]

During this sequence of events as Union forces sought to identify a vulnerable southern flank to attack, Bragg provided insufficient direction when he removed Stewart's division from Buckner's Corps to support Major General Benjamin Franklin Cheatham (see Figure 4). Bragg's orders were so vague that Stewart "rode back to seek clarification, whereupon he discovered how limited Bragg's understanding of the battle really was" as Bragg told Stewart to advance and "be governed by circumstances."[33] These challenges linked to mission orders emanating from Bragg contributed to confusion down the Confederate line that weakened unity of effort among divisions and hindered tempo.

Stewart's Disciplined Initiative and Risk Acceptance

The Confederates' sole success on the 19th was the result of Stewart's individual efforts in the absence of guidance after he became "frustrated with the indecision and lack of communication in the high command" and attacked.[34] Here, Stewart displayed superior disciplined initiative, or "the duty individual subordinates have to exercise [enterprise or leadership] within the constraints of the commander's intent to achieve the desired end state."[35] Throughout the battle, Stewart and his brigade commanders had "not deviate[d] from the published doctrine of the time,"[36] but this changed as Stewart formed his at-

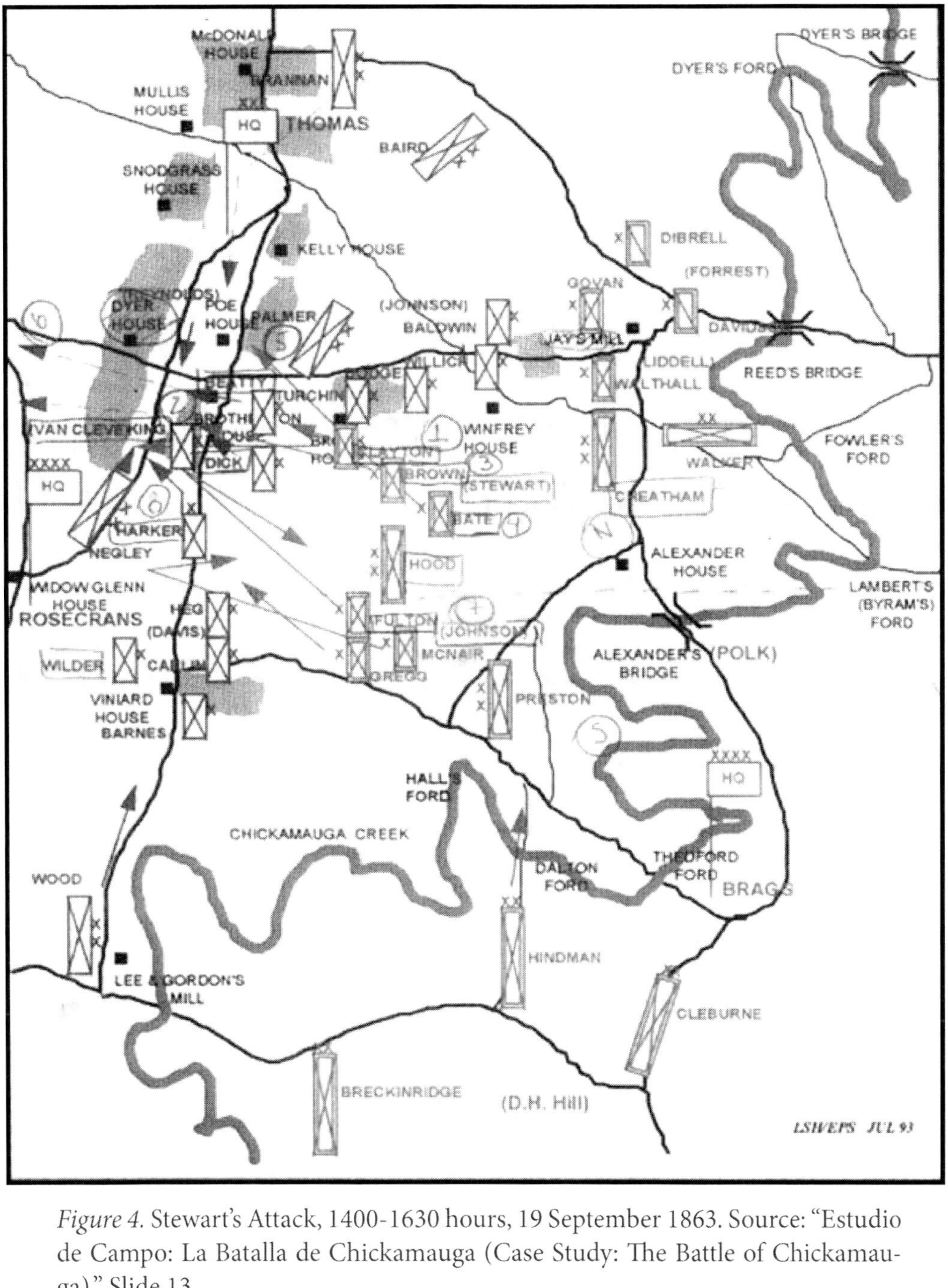

Figure 4. Stewart's Attack, 1400-1630 hours, 19 September 1863. Source: "Estudio de Campo: La Batalla de Chickamauga (Case Study: The Battle of Chickamauga)," Slide 13.

tack formation on the 19th. Rather than the common attack practice of committing successive lines of forces "until they all mingled together into the enemy," Stewart "only committed one brigade at a time," and when that brigade reached its culminating point, he replaced it with a different brigade.[37] Stewart's attack broke the Union's first line and disrupted its second line,[38] which created a breach of the Union line in his sector, but adjacent units failed to exploit this gap (Figure 4).

Stewart accepted risk in this attack, as it carried possibilities of limited support from adjacent units. Furthermore, he decided to lead with his most inexperienced brigade (Clayton's brigade) followed by his most experienced one (Brown's brigade), which carried the possibility of heavy casualties, but proved favorable in terms of audacity and tempo. Brown's brigade was able to psychologically endure witnessing the carnage Clayton's brigade suffered before it and see the wounded pass through its ranks to the rear, whereas a less experienced unit in its position might have broken under such pressure.[39] Stewart understood the likely perils of attacking a prepared enemy with these inexperienced troops, but he recognized the potential payoff for this attack in the midst of a standstill on the battlefield.

Stewart's display of mission command principles set conditions for a critical juncture in the battle whereby the Confederacy could have penetrated the center of the Union line, but it proved indecisive due to inaction on the part of other leaders to capitalize on this opportunity. With nightfall, Stewart ordered his men to lie on their arms through the night with the enemy around 300 yards from Clayton's line. Since Stewart's interaction with Bragg earlier that day, Stewart "had not seen a superior officer all day."[40] In addition to missed opportunities in Stewart's sector following his penetration of Union defenses, this led Stewart to send messengers to the rear to report on the events of that day and seek guidance, which was to hold his position.[41]

Second Day of Battle (20 September 1863)

After 18,000 men had fallen on the 19th between Union and Confederate forces, Rosecrans held a council of war among his subordinate commanders, whereas Bragg called no council of war.[42] Bragg maintained his intent from previous days—to drive the Union army south and away from Chattanooga. The subsequent activity that day at Chickamauga involved Union blunders that permitted Confederate troops to penetrate their line near Rosecrans, which led Rosecrans and one-third of his army to flee from the battlefield.[43] Eventually, Confederate troops confronted the remaining Union forces under Major General George Henry Thomas at Snodgrass Hill. The arrival of the Union army's Reserve Corps enabled most of the remaining Union forces to retreat northward, which ended the Battle of Chickamauga —a costly Confederate victory.

Initial Shared Understanding Under Bragg, Followed by More Ad Hoc Commands

Despite the challenges of the previous day and confusion resulting from Bragg's bypassing corps commanders in ad hoc orders given to division commanders, Bragg achieved shared understanding regarding his scheme of maneuver and initial plans for the 20th. A particularly challenging mission command principle is creating shared understanding of an operational environment, or "an operation's purpose, problems, and approaches to solving problems,"[44] but Bragg's utilization of his corps commanders and simplicity in his attack plan fostered shared understanding. During the evening of the 19th, Bragg had divided his army into two wings: Lieutenant General Leonidas Polk commanding the right wing, and Lieutenant General James Longstreet leading the left wing.[45]

This benefitted Stewart in terms of mission orders, as Longstreet met with him the morning of the 20th and assured Stewart that he would receive guidance directly from Longstreet throughout the day.[46] This would resolve communication issues associated with lack of guidance or correspondence with his higher headquarters the previous day. Bragg directed the attack to begin on the extreme right just after daylight and would "continue down the line by division in succession," so Stewart, like the other division commanders, understood he was to attack when the adjacent unit to his right moved.[47]

Bragg, however, later adjusted these plans and utilized his staff to convey new orders down the line later that morning. Enemy artillery before 10:00 a.m. disrupted some Confederate preparations down the line, including Stewart's division. Bragg and his staff were then observed "dashing along behind Stewart's line"[48] as they disseminated new orders, contradicting the earlier order of successive attacks down the line. One of Bragg's staff officers delivered Bragg's order to Stewart to "advance at once and attack the enemy" and for "every captain to attack"—an order also passed to all other divisions.[49]

Stewart's Disciplined Initiative and Risk Acceptance at Poe Field

On the 20th, Stewart reverted to traditional doctrine with one brigade forward and two back in his attack at Poe Field, fulfilling Bragg's order for an immediate attack along the entire front.[50] The circumstances posed significant danger for Stewart's division, as it would attack an entrenched, prepared defender. Furthermore, his division's flank laid exposed to a bend in the Union line, the attack was an ad hoc response to Bragg's order, the attack overlapped the army's right wing, and units accompanying Stewart were disillusioned with the attack.[51] Despite these challenges, Stewart led his men forward.

As Stewart's division advanced, intervisibility lines from a small hill masked its units as their troops climbed the hill toward the Union line, but they came under devastating fire from a Union battery after cresting the hill (see Figure 5).[52] He succeeded in penetrating the Union's first line of breastworks, but this tactical achievement had little effect on the overall battle in his sector.[53] Heavy smoke from the fighting greatly reduced visibility and caused confusion, but Stewart's subordinate commanders understood his intent to push forward and managed to control their formations. Throughout the afternoon, Stewart and

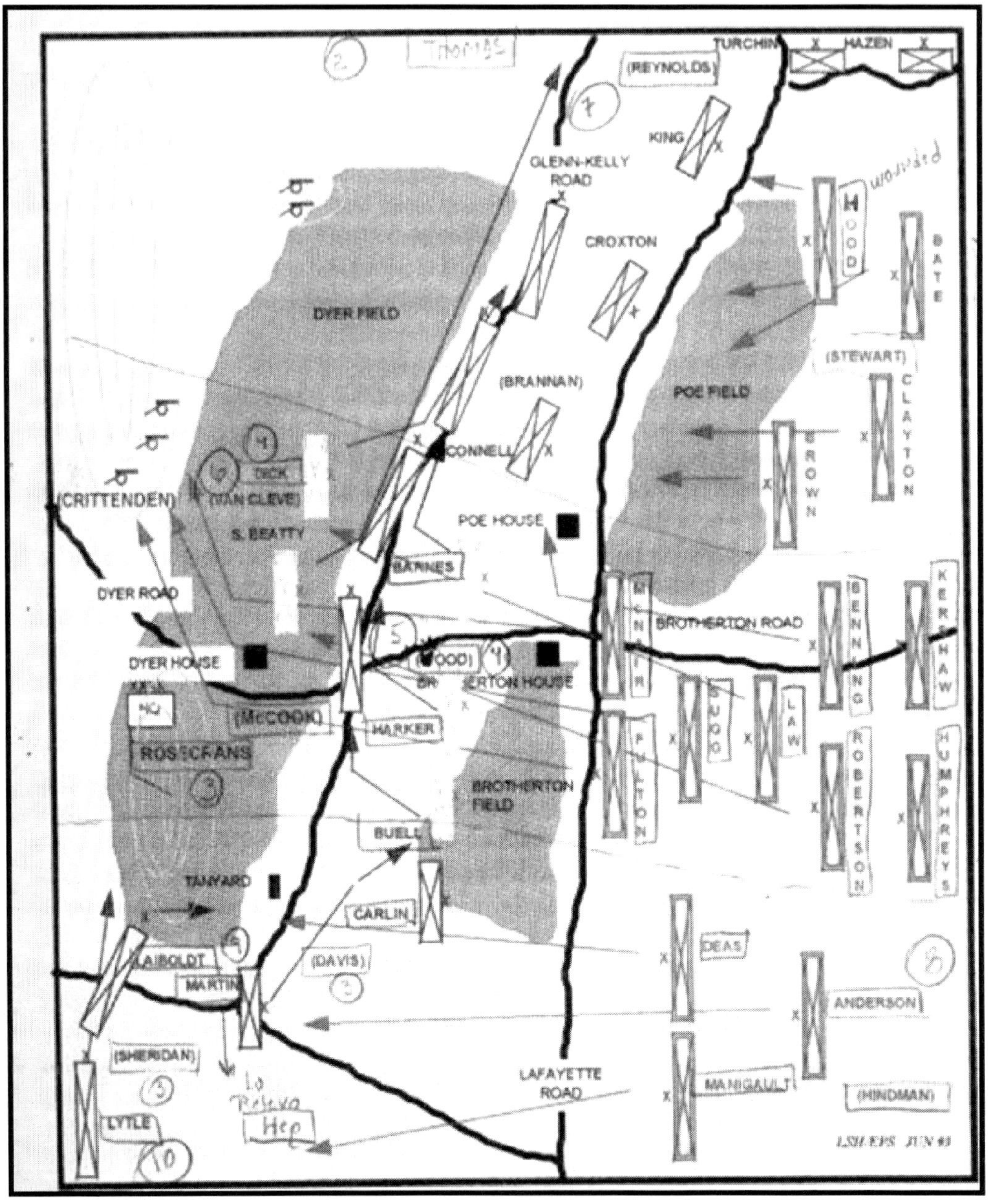

Figure 5. Stewart's Advance at Poe Field, 20 September 1863. Source: "Estudio de Campo: La Batalla de Chickamauga (Case Study: The Battle of Chickamauga)," Slide 20.

Longstreet managed to maintain communications and Stewart received orders that he then executed. His division, working in unity with adjacent units, managed to rout Union forces from the battlefield. Stewart's Eufaula battery, the same unit engaged in the initial engagement on the 18th, claimed to fire the closing shots in the battle as they fired "sixty-nine rounds of shell and canister into fleeing Union forces."[54] Confederate leaders gathered at Stewart's position to oversee the surrender of Federal troops and celebrate their victory.[55]

Aftermath

Under Bragg, the Confederacy secured a victory at Chickamauga, but lost twenty percent of its force in the battle. Its failure to destroy Rosecrans's army, however, permitted Union forces to retreat northward. General Ulysses S. Grant later relieved Rosecrans, and the Union's subsequent attack in November defeated the Army of Tennessee and secured control of the areas around Chattanooga. This enabled Union forces to penetrate the south and destroy factories and disrupt Confederate lines of communication through Major General William Tecumseh Sherman's March to the Sea.

Thus, Bragg's weak mission command contributed to the Confederacy's failure to destroy Rsoecrans's army and permitted the Union to eventually achieve its strategic objective at Chattanooga. After Bragg's defeat by General Ulysses S. Grant in the Battle of Chattanooga in November 1863, he was replaced by Joseph E. Johnston in December 1863. Bragg then became Jefferson Davis's military advisor and commanded the coastal defenses of Wilmington, North Carolina. Bragg's subordinate commanders and historians have blamed his weak mission command for the lackluster results at Chickamauga.

In contrast, Stewart's mission command propelled him to higher offices and led many historians to evaluate him as the best division commander at Chickamauga.[56] His performance led him to steadily take on greater responsibilities and rank. Stewart spent the majority of the war in command of a brigade or a division, but he was "thrust into corps command during the chaos of the Atlanta Campaign."[57] Some aspects of Stewart's mission command abilities at this elevated echelon are inconclusive. For instance, the record is mixed regarding Stewart's culpability for Hood's defeat at Spring Hill. Hood insisted his subordinate generals, including Stewart, failed to follow orders, but others argue such orders were never issued. Veteran D.W. Sanders went further to state that Stewart "was an able and accomplished general" and that had Hood actually ordered Stewart to attack, the result would have been "one of the most brilliant and bloody episodes of the late war."[58] Most evidence of Stewart's mission command after Chickamauga, however, suggests brilliance, such as accounts of his exemplary performance at New Hope Church in May 1864.

Conclusion

At the Battle of Chickamauga, Bragg displayed weak mission command through his incompetence regarding the use of favorable terrain, his vague and ad hoc mission orders, and lack of mutual trust with his subordinate commanders, all of which hindered his army's ability to defeat the Union army; Stewart, on the other hand, displayed strong mission command abilities, most notably in his tactical competence, disciplined initiative, risk acceptance, and the mutual trust among his unit members, all of which enabled his division to secure key ter rain and create opportunities to defeat the Union army. Instances of application or demonstrations of the principles of mission command—competence, mutual trust, shared understanding, commander's intent, mission orders, disciplined initiative, and risk acceptance—reveal significant contrasts between the mission command of Bragg and Stewart in the battle. Had the Army of Tennessee been under the command of a different leader, perhaps the Confederacy would have destroyed the Union army under Rosecrans in early September 1863 or at a place other than Chickamauga. Even during the battle, another operational leader might have capitalized on opportunities presented by the gaps in Union lines created by Stewart's division, but Bragg's poor mission command consistently squandered such opportunities and eventually ceded the strategic key terrain of Chattanooga to the Union.

Bibliography

Alderson, William T. Ed. "The Civil War Diary of Captain James Litton Cooper, September 30, 1861 to January, 1865." *Tennessee Historical Quarterly* 15. 1956. As cited in Barnett, Luke J., III. "Alexander P. Stewart and the Tactical Employment of his Division at the Battle of Chickamauga." Master's thesis, Fort Leavenworth, KS, 1989. https://apps.dtic.mil/sti/pdfs/ADB136021.pdf.

American Battlefield Trust. "Chickamauga: Animated Battle Map." YouTube. June 28, 2019. https://www.youtube.com/watch?v=vlJUuNny9mc&t=538s. Accessed September 15, 2021.

Barnett, Luke J., III. "Alexander P. Stewart and the Tactical Employment of his Division at the Battle of Chickamauga." Master's thesis, Fort Leavenworth, KS, 1989. https://apps.dtic.mil/sti/pdfs/ADB136021.pdf.

Confederate Veteran 3. 1895. As cited in Glaze, Robert Lamar. "Experiencing Defeat, Remembering Victory: The Army of Tennessee in War and Memory, 1861-1930." PhD diss., University of Tennessee, 2016. https://trace.tennessee.edu/cgi/viewcontent.cgi?article=5258&context=utk_graddiss.

Connelly, Thomas L. *Autumn of Glory: The Army of Tennessee, 1862-1865*. Baton Rouge: Louisiana State University Press, 1971. As cited in Glaze, Robert Lamar, "Experiencing Defeat, Remembering Victory: The Army of Tennessee in War and Memory, 1861-1930." PhD diss., University of Tennessee, 2016. https://trace.tennessee.edu/cgi/viewcontent.cgi?article=5258&context=utk_graddiss.

Department of the Army. *Mission Command: Command and Control of Army Forces*, ADP 6-0. Washington, D.C.: Department of the Army, 2019. https://armypubs.

army.mil/epubs/DR_pubs/DR_a/ARN18314-ADP_6-0-000-WEB-3.pdf.

Elliott, Sam Davis. *Soldier of Tennessee: General Alexander P. Stewart and the Civil War in the West*. Baton Rouge, LA: LSU Press, 2004. Kindle Edition.

Glaze, Robert Lamar. "Experiencing Defeat, Remembering Victory: The Army of Tennessee in War and Memory, 1861-1930." PhD diss., University of Tennessee, 2016. https://trace.tennessee.edu/cgi/viewcontent.cgi?article=5258&context=utk_graddiss.

Glazener, A. M. *A. M. Glazener to T. M. Shuford, August 17, 1863*. Letter. From A. M. Glazener Letters, CCNMP. As cited in Elliott, Sam Davis, *Soldier of Tennessee: General Alexander P. Stewart and the Civil War in the West*. Baton Rouge, LA: LSU Press, 2004. Kindle Edition.

Glazener, A. M. *A. M. Glazener to wife, August 20, 26, 1863*. From Glazener Letters. As cited in Elliott, Sam Davis, *Soldier of Tennessee: General Alexander P. Stewart and the Civil War in the West*. Baton Rouge, LA: LSU Press, 2004. Kindle Edition.

Hill, Daniel H. "Chickamauga—The Great Battle of the West." *Battles & Leaders*. As cited in Elliott, Sam Davis, *Soldier of Tennessee: General Alexander P. Stewart and the Civil War in the West*. Baton Rouge, LA: LSU Press, 2004. Kindle Edition.

Kirby, Jno. A. *Jno. A. Kirby to "Pa," August 26, 1863*. Confederate States Archives, Officers and Soldiers' Miscellaneous Letters, DU. As cited in Elliott, Sam Davis, *Soldier of Tennessee: General Alexander P. Stewart and the Civil War in the West*. Baton Rouge, LA: LSU Press, 2004. Kindle Edition.

Lindsley, John Berrian. Ed. *The Military Annals of Tennessee: Confederate*. Nashville: J. M. Lindsley, 1886. As cited in Elliott, Sam Davis, *Soldier of Tennessee: General Alexander P. Stewart and the Civil War in the West*. Baton Rouge, LA: LSU Press, 2004. Kindle Edition.

McMurray, W. J. "The Gap of Death at Chickamauga." *Confederate Veteran* 2 (November 1894). As cited in Sam Davis Elliott, *Soldier of Tennessee: General Alexander P. Stewart and the Civil War in the West*. Baton Rouge, LA: LSU Press, 2004. Kindle Edition.

Ogden, James, III. Presentation for Western Hemisphere Institute for Security Cooperation (WHINSEC) students. Chickamauga, GA. September 16, 2021.

Ogden, James, III. Presentation for Western Hemisphere Institute for Security Cooperation (WHINSEC) students. Chickamauga, GA. September 17, 2021.

Powell, David A. "The Battle of Chickamauga." Developed for Virginia Center for Civil War Studies at Virginia Tech, Blacksburg, VA, 2010. https://www.essentialcivilwarcurriculum.com/the-battle-of-chickamauga.html. Accessed September 27, 2021.

Ridley, Bromfield L. *Battles and Sketches of the Army of Tennessee*. Mexico, Missouri: Missouri Printing and Publishing Co., 1906. Reprint: Dayton, Ohio: Morningside Bookshop, 1978. As cited in Barnett, Luke J., III, "Alexander P. Stewart and the Tactical Employment of his Division at the Battle of Chickamauga." Master's thesis, Fort Leavenworth, KS, 1989. https://apps.dtic.mil/sti/pdfs/ADB136021.pdf.

Southern Bivouac 3. 1884. As cited in Glaze, Robert Lamar. "Experiencing Defeat, Remembering Victory: The Army of Tennessee in War and Memory, 1861-1930." PhD diss., University of Tennessee, 2016. https://trace.tennessee.edu/cgi/viewcontent.cgi?article=5258&context=utk_graddiss.

Stephenson, Philip Dangerfield. *The Civil War Memoir of Philip Dangerfield Stephenson, D. D.: Private, Company K, 13th Arkansas Volunteer Infantry and Loader, Piece No. 4, 5th Company, Washington Artillery, Army of Tennessee, CSA*. Ed. Nathaniel Cheairs Hughes, Jr. Conway: University of Central Arkansas Press, 1995. As cited in Glaze, Robert Lamar, "Experiencing Defeat, Remembering Victory: The Army of Tennessee in War and Memory, 1861-1930." PhD diss., University of Tennessee, 2016. https://trace.tennessee.edu/cgi/viewcontent.cgi?article=5258&context=utk_graddiss.

Stinson, Robert Daniel. "Braxton Bragg on Leadership: A Historical Case Study on Leader-Member Exchange in the Army of Tennessee." EdD diss., University of Tennessee at Chattanooga, 2013.

"Tennessee C.S.A. Infantry." *The Historical Marker Database*. https://www.hmdb.org/m.asp?m=62628. Accessed October 10, 2021.

U.S. Army Command and Staff College. "The American Civil War: Alexander P. Stewart and the Tactical Employment of His Division at the Battle of Chickamauga." Damascus, MD: Penny Hill Press, 2015.

U.S. War Department. *War of the Rebellion: A Compilation of the Official Records of the Union and the Confederate Armies*. 70 Vols. in 128 parts. Washington, D.C.: Government Printing Office, 1880-1901.

Notes

1 Robert Daniel Stinson, "Braxton Bragg on Leadership: A Historical Case Study on Leader-Member Exchange in the Army of Tennessee," dissertation written for the University of Tennessee at Chattanooga, Chattanooga, TN, May 2013, 4.

2 W. J. McMurray, "The Gap of Death at Chickamauga," *Confederate Veteran* 2 (November 1894), 330, as cited by Elliott, *Soldier of Tennessee* (Baton Rouge, LA: LSU Press, 2004), Kindle Edition, location 3850. In the same reference, Elliott also wrote that "Stewart's feelings toward a monument to his memory at a time he was still living (in 1894) are unknown."

3 On this, see Elliott, *Soldier of Tennessee*, location 1399; "Tennessee C.S.A. Infantry," *The Historical Marker Database*, https://www.hmdb.org/m.asp?m=62628.

4 Elliott, *Soldier of Tennessee*, location 1406.

5 Department of the Army, *Mission Command: Command and Control of Army Forces*, ADP 6-0 (Washington, D.C.: Department, of the Army, 2019), https://armypubs.army.mil/epubs/DR_pubs/DR_a/ARN18314-ADP_6-0-000-WEB-3.pdf, Glossary-3.

6 U.S. Army Command and Staff College, "The American Civil War: Alexander P. Stewart and the Tactical Employment of His Division at the Battle of Chickamauga" (Damascus, MD: Inc. Penny Hill Press, 2015), 59-60.

7 Department of the Army, *Mission Command*, para. 1-27.

8 Private A. M. Glazener of the 18th Alabama penned in a letter to a friend dated 17 August: "This army is in a bad condition. The men all think we are whipt all out of heart they don't put much confidence in this general, Bragg." On this, see A. M. Glazener to T. M. Shuford, August 17, 1863 (typed copy), A. M. Glazener Letters, CCNMP; see Glazener to wife, August 20, 26, 1863, Glazener Letters; Jno. A. Kirby to "Pa," August 26, 1863, Confederate States Archives, Officers and Soldiers' Miscellaneous Letters, DU, as cited by Elliott, locations 1166 and 3827.

9 It is likely that forces in vicinity of the modern Point Park, TN, just on the southwest side of Chattanooga, would have been able to observe indications of a large Union force advancing through Lookout Valley, based on the author's visit to the site on 16 September 2021. This terrain feature affords observation over vast swaths of the Tennessee River and valley westward.

10 On this, see Hill, "Chickamauga," *B&L* 3:641-42, as noted in Elliott, *Soldier of Tennessee*, location 3833 (footnote 13).

11 Ogden, presentation, 16 September 2021.

12 Elliott, *Soldier of Tennessee*, location 1209.

13 U.S. Army Command and Staff College, "The American Civil War," 32-33.

14 On this, see Hill, "Chickamauga," *B&L* 3:641-42, as noted in Elliott, *Soldier of Tennessee*, location 3833 (footnote 13).

15 Lindsley, ed., *Military Annals*, 81, as cited in Elliott, *Soldier of Tennessee*, location 1196.

16 Stephenson, *Civil War Memoir*, 212, as cited in Robert Lamar Glaze, "Experiencing Defeat, Remembering Victory: The Army of Tennessee in War and Memory, 1861-1930," doctoral dissertation written for the University of Tennessee, Knoxville, August 2016, https://trace.tennessee.edu/cgi/viewcontent.cgi?article=5258&context=utk_graddiss, 81.

17 *Confederate Veteran* 3 (1895): 260-61, as cited in Glaze, "Experiencing Defeat, Remembering Victory," 81.

18 *Confederate Veteran* 3 (1895): 260-61, as cited in Glaze, "Experiencing Defeat, Remembering Victory," 81.

19 Elliott, *Soldier of Tennessee*, location 1151.

20 Ibid., location 1171.

21 Ogden, presentation, 16 September 2021.

22 Department of the Army, *Mission Command*, para. 1-45.

23 Ogden, presentation, 16 September 2021.

24 *The War of the Rebellion: A Compilation of the Official Records of the Union and Confederate Armies*, 70 vols. in 128 parts (Washington D. C.: Government Printing Office, 1880-1901), Series 1, vol. XXX, Pt. 2, 357, 361, 382, 384, and 394; Bromfield L. Ridley, *Battles and Sketches of the Army of Tennessee* (Mexico. Missouri: Missouri Printing & Publishing Co., 1906; reprint, Dayton, Ohio: Morningside Bookshop, 1978), 208-209 (references are to reprint edition); as cited in Luke J. Barnett III, "Alexander P. Stewart and the Tactical Employment of his Division at the Battle of Chickamauga," master's thesis presented at Fort Leavenworth, KS, 1989, https://apps.dtic.mil/sti/pdfs/ADB136021.pdf, 85.

25 Barnett, "Alexander P. Stewart," 88.

26 William T. Alderson, ed., "The Civil War Diary of Captain James Litton Cooper, September 30, 1861 to January, 1865," Tennessee Historical Quarterly 15 (1956): 157; Ridley, *Battles & Sketches*, 209; *The War of the Rebellion*, Series 1, vol. XXX, Pt. 2, 361, 384, 394, and 401; as cited in Barnett, "Alexander P. Stewart," 86 and 88.

27 Barnett, "Alexander P. Stewart," 89.

28 Elliott, *Soldier of Tennessee*, location 1255.

29 James Ogden III, presentation for Western Hemisphere Institute for Security Cooperation (WHINSEC) students, Chickamauga, GA, 17 September 2021.

30 Department of the Army, *Mission Command*, para. 1-52 and 1-54.

31 Ibid., para. 1-56.

32 David A. Powell, "The Battle of Chickamauga," developed for Virginia Center for Civil War Studies at Virginia Tech, Blacksburg, VA, 2010, https://www.essentialcivilwarcurriculum.com/the-battle-of-chickamauga.html.

33 United States War Department, *The War Of The Rebellion*, Series I, Volume 30, part 2, 361, as cited in Powell, "The Battle of Chickamauga."

34 Thomas L. Connelly, *Autumn of Glory: The Army of Tennessee, 1862-1865* (Baton Rouge: Lousiana State University Press, 1971), 201-203, as cited and summarized in Glaze, "Experiencing Defeat, Remembering Victory," 25.

35 Department of the Army, *Mission Command*, para. 1-59.

36 Barnett, "Alexander P. Stewart," 147.

37 Ibid., 146-147.

38 Ibid., 149-150.

39 Ibid., 149.

40 U.S. Army Command and Staff College, "The American Civil War," 106.

41 *The War of the Rebellion*, Series 1, vol. XXX, Pt. 2, 363, 371, 405, 376, 382, 384, 393-94, and 397, as cited in U.S. Army Command and Staff College, "The American Civil War," 106.

42 American Battlefield Trust, "Chickamauga: Animated Battle Map," YouTube, 28 June 2019, https://www.youtube.com/watch?v=vlJUuNny9mc&t=538s, 10:50.

43 American Battlefield Trust, "Chickamauga: Animated Battle Map," 16:17.

44 Department of the Army, *Mission Command*, para. 1-36.

45 U.S. Army Command and Staff College, "The American Civil War," 117.

46 Ibid., 117.

47 Ibid., 117-118.

48 Ibid., 120.

49 Ibid., 120-122.

50 Barnett, "Alexander P. Stewart," 148.

51 Ibid., 147-148.

52 U.S. Army Command and Staff College, "The American Civil War," 123.

53 Barnett, "Alexander P. Stewart," 147 and 150.

54 U.S. Army Command and Staff College, "The American Civil War," 133.

55 Ibid., 134.

56 Ibid., 49.

57 Glaze, "Experiencing Defeat, Remembering Victory," 80.

58 *Southern Bivouac* 3 (1884): 365, as cited in Glaze, "Experiencing Defeat, Remembering Victory," 80.

Evaluating Tiberius' Character According to Seneca's Father/Son—Master/Slave Paradigm *On Mercy*

By Mary Jo Davies
American Public University

ABSTRACT

As a slave-master, a Roman father ruled under the guise of *pater potesta* (father's power), which allowed him to kill his slaves without compassion. However, Roman society expected a father to show his own children more *clementia* (mercy) and only kill them after more tempered methods of discipline had failed. As *pater patriae* (father of the country), a Roman emperor had these same authoritative privileges over the entire Roman population. Since the reigns of Tiberius and Caligula had degenerated into ruthless despotism, Seneca (4 BC—65 AD) wrote *On Mercy* to warn the newly appointed young emperor Nero (r. 54 AD—68 AD) that as *pater patriae* he should treat his free subjects according to the father/son model rather than the master/slave model. By showing his free subjects more mercy, he would win public esteem and ensure a successful reign. Although all emperors exhibited some measure of ruthlessness toward their subjects, Rome's understanding of good and bad imperial behavior in the first century AD, specifically in accordance with Seneca's father/son—master/slave model, demonstrates why Rome considered Tiberius a bad emperor despite his many accomplishments.

Keywords: *pater potesta, pater patriae, clementia*, father/son, master/slave, *autorictas*, principate, one-man rule, *princeps, maiestas*

Evaluación del carácter de Tiberio según el paradigma padre/hijo-amo/esclavo de Séneca sobre la misericordia

RESUMEN

Como amo de esclavos, un padre romano gobernaba bajo la apariencia de *pater potesta* (poder del padre), lo que le permitía matar a sus esclavos sin compasión. Sin embargo, la sociedad romana

 doi: 10.18278/sshj.10.3.7

esperaba que un padre mostrara a sus propios hijos más *clementia* (misericordia) y solo los matara después de que los métodos de disciplina más templados hubieran fallado. Como *pater patriae* (padre de la patria), un emperador romano tenía estos mismos privilegios de autoridad sobre toda la población romana. Dado que los reinados de Tiberio y Calígula habían degenerado en un despotismo despiadado, Séneca (4 a. C.—65 d. C.) escribió Sobre la misericordia para advertir al recién nombrado joven emperador Nerón (r. 54 d. C.—68 d. C.) que debía tratar a sus libres como *pater patriae*. sujetos según el modelo padre/hijo en lugar del modelo amo/esclavo. Al mostrar más misericordia a sus súbditos libres, ganaría la estima pública y aseguraría un reinado exitoso. Aunque todos los emperadores exhibieron cierta medida de crueldad hacia sus súbditos, la comprensión de Roma del buen y mal comportamiento imperial en el siglo I d. C., específicamente de acuerdo con el modelo padre/hijo-amo/esclavo de Séneca, demuestra por qué Roma consideraba a Tiberio un mal emperador a pesar de su muchos logros.

Palabras clave: *pater potesta, pater patriae, clementia,* padre/hijo, amo/esclavo, *autorictas, principado,* gobierno de un solo hombre, *princeps, maiestas*

通过塞内卡的父/子-主人/奴隶仁慈范式评价提贝里乌斯

摘要

作为奴隶主，古罗马家父（Roman father）在家父绝对权威（pater potesta）这一表现形式下进行统治，后者允许其无情地杀害奴隶。不过，古罗马社会期望父亲对其子女展现更多仁慈（clementia），并且仅在更为缓和的训诫方式失败后才能杀死孩子。作为国父（pater patriae），罗马皇帝对整个罗马公民拥有同样的绝对权威。自提贝里乌斯和卡利古拉堕落为无情的暴君后，塞内卡（4 BC—65 AD）撰写了《论仁慈》警告新任命的年轻君主尼禄（54 AD—68 AD）：作为国父他应以父/子的模式对待他的自由臣民，而不是根据主人/奴隶模式。通过对自由臣民施以更多仁慈，他将赢得民众的尊敬并确保成功的统治。尽管所有君主都对其臣民展现一定程度的无情，但公元1世纪罗马对君主善恶行为的理解（即根据塞内卡的父/子-主人/奴隶模式）证明了为何罗马将提贝里乌斯视为残忍君主，尽管其取得了许多成就。

关键词：家父绝对权威，国父，仁慈，父/子，主人/奴隶，权威，元首制，一人统治，第一公民，叛国（maiestas）

> For kings, however, the surest way to security is through gentleness, since frequent punishment, while it crushes hatred in a few, arouses it in everyone.[1]

In ancient Rome, the privileges a man held over his family (including his slaves) were exceedingly authoritative. As a slave-master he ruled under the guise of *pater potesta* (father's power), which allowed him to kill his slaves without pity or compassion. Roman society expected fathers to show their own children *clementia* (mercy) and only resort to execution after more tempered methods of discipline had failed. As *pater patriae* (father of the country), a Roman emperor had these same privileges over the entire Roman population. However, society expected even an emperor to show his free-born and aristocratic subjects more mercy than Rome's slave population. But this was not always the case. The ruthlessness some emperors exhibited often did not take social status into consideration.

Roman philosopher Seneca (4 BC–65 AD) had seen the very best in Augustus (*r.* 27 BC–14 AD), the first emperor of Rome, and the very worst in his immediate successors, Tiberius (*r.* 14 AD–37 AD) and Caligula (*r.* 37 AD–41 AD). The reigns of Tiberius and Caligula had degenerated into ruthless despotism. As advisor to the newly appointed young emperor Nero (*r.* 54 AD–68 AD), Seneca wrote *On Mercy* as a means to prevent this kind of rule from becoming standard in the newly established principate. Nero was fearful that his thirteen-year-old stepbrother, Brittanicus, would one day make himself emperor. Brittanicus was the biological son of Claudius (*r.* 41 AD–54 AD), the fourth emperor of Rome and true heir to his father's sovereignty, but he was much too young to take on the role of emperor at the time of his father's death. This paved the way for Claudius's stepson, Nero, to become the fifth emperor of Rome. To safeguard his sovereignty, Nero had Brittanicus murdered.[2]

Although Seneca flatters Nero in *On Mercy*, his intention is to advise the young ruler that to win the trust of the populace he should adopt the more compassionate father/son model rather than the coldblooded master/slave model toward his free-born and aristocratic subjects. Since Tiberius's reign had taken a dark, despotic turn, he was clearly one of the emperors that compelled Seneca to write *On Mercy* as a warning to Nero. Although all emperors exhibited some measure of ruthlessness toward their subjects, Rome's understanding of good and bad imperial behavior in the first century AD, specifically in accordance with Seneca's father/son—master/slave model, demonstrates why Rome considered Tiberius a bad emperor despite his many accomplishments.

Seneca's theory was based on Roman historian Quintus Curtius's (*c.* first century AD) account of Hermolaus in the *History of the Wars of Alexander.* Hermolaus was one of many noble adolescent youths whose family entrusted as attendant to Alexander the Great (*r.* 336 BC–323 BC).[3] Among the duties of a royal attendant was that of accompanying the king on hunting expeditions. On one occasion Hermolaus slayed a wild boar without giving Alexander the royal honor of first strike. Infuriated by this breach of protocol, Alexander had Hermolaus punished with stripes. Hermolaus deemed it humiliating that the king should punish him in the same manner as he would a slave. His punishment led to a failed attempt, with fellow conspirators, to assassinate the king. For this attempt, Alexander had Hermolaus stoned to death.[4] Curtius's account follows Alexander's psychological evolution from a brilliant young conqueror to a ruthless leader who often made use of the master/slave model to punish his free-born and aristocratic subjects.

The father/son relationship appears quite frequently in Julio-Claudian texts as a more suitable model for an emperor to adopt toward his free subjects.[5] One of these texts is Seneca's *On Mercy*. While Seneca believed that showing restraint toward slaves was praiseworthy, he also believed that an emperor should "not abuse the free, the free-born, the well-born as though they were chattels."[6] The intent behind this father/son approach to leadership was to prevent despotism from happening again now that Rome had returned to one-man rule.

Rome had started out as a monarchy circa 750 BC. Nearly two-hundred-fifty years later its monarchs had increasingly turned to despotism to rule the populace. In 509 BC, Rome abolished the monarchy and established a Republic with a mixed constitution in an effort to prevent the kind of autocracy that had developed under their oppressive kings. The Republic lasted nearly 500 years when the civil war brought it to an end—at which time Rome had returned to one-man rule by establishing a principate in 27 BC.

Seneca believed that an emperor should use his authority judiciously to improve society and garner the respect of his subjects rather than to instill fear. A despot, on the other hand, uses his power to punish and kill any member of society indiscriminately, unreasonably, and often based on anger and suspicion. In *On Mercy*, Seneca cautions Nero,

> Mercy, then, enhances not only a ruler's honour, but his safety. The glory of the empire, it is at the same time its surest protection. For what has allowed kings to grow old and bequeath their kingdom to their children and grandchildren? Why is the power of tyrants accursed and short-lived? What is the difference between a king and a tyrant? – after all, their show of fortune and their power are the same. It is simply that tyrants act savagely for their pleasure, whereas kings do so only for a reason and out of necessity.[7]

Ancient accounts have balanced the brilliant leadership of Augustus with his ruthlessness, the latter which he exhibited during his rise to power, but mitigated throughout his reign as emperor. Yet history has always classified him as a good emperor. Sources have also recorded Tiberius's many accom-

Figure 1. *Nero and Seneca,* by Eduardo Barrón González (1904). Museo del Prado. Licensed under the Creative Commons.

plishments, yet they ultimately deemed him to be a bad emperor *because* of his ruthlessness. Assessing Tiberius's character therefore requires fully understanding the perceptions of good and bad imperial behavior that fall within the parameters of Roman culture in the first century AD.

In his article "The *Diritas* of Tiberius," historian Kenneth Scott recognized the emperor's disposition as one that "could never win popular favor" because of his cynicism and brutality.[8] However, his analysis lacked the depth needed to interpret Tiberius's personality according to ancient cultural standards. He merely tried to evaluate the charge of *diritas* that Tiberius's stepfather, Augustus, brought against him by finding proof of such cruelty in the primary sources.[9] Augustus knew that while Tiberius enjoyed a positive military reputation in his early career, he also possessed a dark side—an unlikeable, awkward, and arrogant disposition.

Despite Tiberius's unpleasant temperament, the first five years of his reign were quite successful. A cultured, intelligent man, he was very capable and experienced not only in military affairs, but in his administration and diplomacy.[10] In *The Life of the Twelve Caesars,* Suetonius (69 AD–122 AD) commented rather favorably on his foreign policy

(Top) ***Figure 2.*** "Tiberius coin," British Museum R.6427. (Bottom) ***Figure 3.*** "Tiberius coin," British Museum 1938,0510.118. Both images © The Trustees of the British Museum Shared under a Creative Commons, Attribution-NonCommercial-ShareAlike 4.0 International license (CC BY-NC-SA 4.0)

saying, "[h]e preferred to employ reprimands and threats, rather than force, against foreign kings suspected of disaffection."[11]

From the beginning of his reign, Tiberius was inclined toward frugality such as reducing the cost of games and shows and slashing pay for actors, among many other things. To encourage thriftiness among the masses he often served left-over meats at his own formal dinners.[12] The reverse of his coins frequently displayed the double cornucopia (see Figure 2), a symbol of bounty, which seems to reflect what his measures generated.

Tiberius also disapproved of flattery and excessive honors and refused the kind of emperor worship which Roman society had accorded his predecessor, Augustus. In this vein, he also refused to take the honorary title of *Augustus* since he wished to remain at the same human level as everyone else. This certainly seemed to detach his leadership from the terrible memory of Rome's despotic kings. The only time he allowed the use of the title *Augustus* was on inscriptions, on coins (see Figure 3) and in formal letters.[13] Dio alleged that "he was so democratic in all circumstances alike, that he would not permit any special observance to be made of his birthday."[14]

All these qualities might seem to justify emphasizing Tiberius's historical legacy as a good leader and downplaying his ruthlessness, just as history did with Augustus. However, while Augustus did have people executed throughout his reign, it was his overall predisposition toward imperial *autorictas* that made the difference between a father/son and a master/slave relationship. Under the principate, Augustus's brutal, pre-imperial rise to power did not matter anymore. Once he donned the robe of *princeps* he had adopted a father/son relationship with the populace. Seneca summed up Augustus's character best in the following passage,

> [W]e acknowledge that Augustus was a good prince, richly deserving the name of 'parent,' for the simple reason that the insults offered to him, which princes usually find more bitter than injuries, were never avenged with cruelty, that he

Figure 4. Bronze statue of the Roman emperor Tiberius with head veiled. Found in the theater in Herculaneum. Licensed under the Creative Commons.

Figure 5. *The Death of Tiberius* by Jean Paul Laurens, 1864; Musée Paul-Dupuy. Licensing: this media file is in the public domain.

> smiled at the abuse directed at him, that he appeared to be suffering punishment when inflicting it, that, so far from killing those whom he had condemned for adultery with his daughter, he sent them away for their safety.[15]

Unlike Augustus, Tiberius felt uncomfortable assuming the role of *princeps*. He had been very successful in his military role prior to becoming emperor. In the late 20s BC, his successful diplomatic and military activities allowed for the recovery of the legionary standards, which Roman generals, Crassus and Decidius Saxa had lost to Parthia in 53 BC in 36 BC respectively. Tiberius's role in this victorious mission ultimately ensured Rome's superiority over Armenia and Parthia.

For the next few decades, the battlefields of Europe kept Tiberius busy and distinguished him as an efficient commander, yet he remained an awkward, rather haughty man. His morose personality may not have made him very popular in Rome, but people nevertheless respected him because of his productive military role. Tiberius liked the fact that the battlefields kept him far away from the politics of Rome and gave him the self-assurance he needed to be successful.[16]

However, back home Augustus was in the process of selecting his successor. He knew the one thing Romans hated most was dynastic rule for fear that the Empire would turn into another oppressive monarchy like that of Rome's pre-Republic years, yet he did not want to leave succession to the next best man fearing that it could lead to conflict and another civil war such as happened with the Republic. To avoid suspicion of dynastic succession he was careful to name an heir by way of indirect designation—that is, succession by way of marriage into the Julio-Claudian line. What helped was the fact that Augustus had no son of his own.

Tiberius had not been first choice as heir, but every time Augustus chose a successor, the man he appointed died, eventually making Tiberius the last choice. To make him an official dynastic member of the Julio-Claudian line, Augustus forced Tiberius to divorce his wife Vipsania, with whom he was very happy. Augustus then had his own daughter, Julia—widow of Agrippa—marry Tiberius.[17] Tiberius's fate as future emperor, made possible by his mandatory divorce from Vipsania, caused him much misery. Julia's promiscuity further exacerbated matters. When he received tribunician powers in 6 BC, he went into a self-imposed exile in Rhodes. This forced Augustus to rely on Julia's two young sons from her previous marriage to Agrippa: fourteen-year old Gaius and eleven-year old Lucius. Since Augustus's health was never very good, he hoped to survive until at least one of the boys was old enough to take on the role of emperor. Augustus suffered many health issues, among them ringworm, abscessed liver, and influenza. His worst illness came very early in his reign as emperor in the year 23 B.C., when he almost lost his life.[18]

Augustus survived and went on to live many more years, but Gaius and Lucius both died on official missions to Spain—Lucius in 2 AD and Gaius in 4 AD. This obliged Augustus to ultimately give Tiberius no choice and named him his official successor. Tiberius resigned himself to his destiny. Nine years later, in 13 AD, he received imperial powers equal to those of Augustus. The following year, in 14 AD, Augustus died, making Tiberius emperor of Rome. Tiberius knew instantly that he needed to win the favor of the Roman people.[19]

In *Roman History,* Cassius Dio (155 AD–235 AD) established a narrative timeline when he determined that up until his nephew Germanicus's death in 19 AD, "Tiberius had done a great many excellent things and had made but few errors."[20] But Germanicus's military accomplishments made him very popular in Rome. Despite imperial guidance from Tiberius in quelling a German uprising in 17 AD, it was mostly Germanicus's actions in response to his uncle's guidance that earned the adoration of the people and garnered him a triumphal celebration on his return to Rome. In the *Annals,* Tacitus (56 AD–120 AD) dedicated a great deal more space to Germanicus in his account of Tiberius than of the emperor himself. Germanicus appeared to have all the well-respected qualities that Tiberius lacked, but he began to assume too much self-directed power for the emperor's liking.

In 19 AD, Germanicus died under mysterious circumstances—seemingly by poison. His death provoked much speculation in Rome that Piso, then governor of Syria, acted under Tiberius's orders to dispose of Germanicus for fear that Germanicus was planning to seize the sovereignty from his uncle. Although Tiberius did not enjoy being emperor, he no doubt feared the shame of defeat, particularly from his much younger, publicly beloved nephew.

Tacitus states that on the day of Germanicus's burial, the city was, "desolate in its silence" and "distracted by lamentations."[21] The public display of grief and adulation toward Germanicus irritated Tiberius. He would not even allow his nephew to be buried with a gold shield. Rather, he insisted on commissioning "a normal one, identical to the rest."[22] It is because of the dichotomy of Tiberius's behavior (democratic yet vengeful) that he appeared deceptive to Dio, who states, "[p]erhaps he had been at heart from the first what he later showed himself to be, and had been merely shamming while Germanicus was alive, because he saw his rival lying in wait for the sovereignty."[23] Despite the suspicions, they were never able to prove that Tiberius commissioned the death of his nephew.

Whether or not the rumors that Tiberius had a hand in Germanicus's death are true, Dio attributes Tiberius's subsequent degeneration into paranoid despotism to Germanicus's death, stating that "he changed to precisely the reverse of his previous conduct, which had included much that was good."[24] Perhaps in fear of a conspiracy against his own life (the kind of conspiracy that Seneca attributed to a reaction to despotism), Tiberius reinstated the trials of *maiestas* (treason). His increasing distrustfulness dragged his administration through the mud. When his son, Drusus, whom he had been grooming for succession, mysteriously died, Tiberius adopted much more severe tactics for accusations of treason. Since the crime of treason was never well-defined in ancient Rome, such charges, under Tiberius, eventually came to encompass the ruin of men who at most had been discourteous to the emperor.[25]

Tiberius's trials of *maiestas* alone made him a bad ruler according to Roman society in the first century AD, but when considering the father/son—master/slave model, as Seneca expressed in *On Mercy*, it is essential to underscore that Tiberius was not bad because he killed many people. He was bad because, unlike Augustus, he often killed indiscriminately rather than with the benefit of Roman society in mind. Suetonius mentions that Tiberius's intent on improving public morality was, in truth, his cruel way "to gratify his own savage disposition."[26]

While the strength of the Empire did not really suffer from Tiberius's ruthless character, he never did win acceptance from the Senate or the people;[27] not even during the financial crisis of 33 AD, when Tiberius manufactured and provisioned large amounts of cash and endowed the public treasury with 100,000,000 sesterces.[28] His many accomplishments, both early and late in his reign were not enough to cancel his bad deeds. His effort to win public acceptance by rejecting the title of *Augustus* might have seemed egalitarian, even to some of his contemporaries, but as *pater patriae* he was supposed to rule as a "father" to avoid the kind of cruelty that had corrupted Rome's monarchy. This is precisely the reason why the more merciful father/son relationship became so vital in the new principate. But Tiberius knew he had big shoes to fill when Augustus died. Perhaps he was aware that his own naturally arrogant and awkward personality did not make him a likable person and that this was the reason why he turned to sycophantic methods to influence public acceptance not knowing that his successful administrative measures might have been enough to safeguard his sovereignty and his heritage. Nevertheless, it was mostly because of his increasingly ruthless behavior and his paranoid nature, particularly in his treatment of Germanicus and the trials of *maiestas*, that Rome learned to distrust him.

On writing of Tiberius's death in 37 AD at the age of 77, Suetonius states that some people ran about shouting, "[a]way with Tiberius to the Tiber."[29] Tacitus records that Macro, a *prefect* of the *Praetorian Guard*, smothered Tiberius with a pile of clothing.[30]

Had Tiberius died of natural causes before his reign took a dark turn, history might have hailed him as an excellent ruler, but when he adopted a master/slave relationship with the populace, he alienated the people he was supposed to protect and guide and, in the process, destroyed his own legacy.[31]

Bibliography

Primary Sources

Dio, Cassius. *Roman History*. Translated by Ernest Cary, Ph.D. Vol. 7. New York: Harvard University Press, 1924. https://archive.org/details/L175CassiusDioCocceianusRomanHistoryVII5660/page/n171/mode/2up?q=maiestas

Curtius, Quintus. *History of the Wars of Alexander.* Translated by John Digby Esq. London: A. Millar, 1747. https://archive.org/details/quintuscurtiush00freigoog

Seneca. *Moral and Political Essays*. Edited and translated by John M. Cooper and J. F. Procopé. Cambridge: Cambridge University Press, 1995.

Tacitus, Cornelius. *Annals*. The Perseus Catalog. http://www.perseus.tufts.edu/hopper/text?doc=Perseus:text:1999.02.0078

Tranquillus, C. Suetonius. *The Lives of the Twelve Caesars.* Translated by Alexander Thomson. M.D. London: George Bell & Sons, 1893. https://archive.org/details/livestwelvecaes00foregoog/page/n6/mode/2up

Secondary Sources

Boatwright, Mary T., et al. *The Romans: From Village to Empire.* 2nd ed. New York: Oxford University Press, 2004.

Bunson, Matthew. *A Dictionary of the Roman Empire.* New York: Oxford University Press, 1995.

Cary, M. and H. H. Scullard. *A History of Rome.* 3rd ed. New York: Palgrave Macmillan, 1975.

Le Glay, Marcel., et al. *A History of Rome.* 4th ed. Malden: Wiley-Blackwell Publishing Company, 2009.

Roller, Matthew. *Constructing Autocracy: Aristocrats and Emperors in Julio-Claudian Rome.* Princeton: Princeton University Press, 2001.

Scott, Kenneth. "Tiberius's Refusal of the Title "Augustus." *Classical Philology* 27, no. 1 (1932): 43-50.

Scott, Kenneth. "The *Diritas* of Tiberius." *The American Journal of Philology* 53, no. 2 (1932): 139-151.

Shotter, David. *Tiberius Caesar.* 2nd ed. New York: Routledge/Taylor & Francis Group, 2004.

Thornton, M. K. and R. L. Thornton. "The Financial Crisis of A.D. 33: A Keynesian Depression?" *The Journal of Economic History* 50, no. 3 (1990): 655-662.

Notes

1 Seneca, *On Mercy* 1.8; in *Moral and Political Essays* ed. and trans John M. Cooper and J. F. Procopé (Cambridge: Cambridge University Press, 1995).

2 Claudius died seemingly by poison. Some ancient sources differ in their accounts of who administered the poison, while others claim that Claudius's death by poison was merely a rumor. Whatever the case, succession became an issue. Claudius's wife, Agrippina the Younger, favored her own biological son, Nero, from her previous marriage to Gnaeus Domitius Ahenobarbus. Claudius, on the other hand, had been grooming Brittanicus, his own biological son from *his* previous marriage to Valeria Messalina.

3 Matthew B. Roller, *Constructing Autocracy: Aristocrats and Emperors in Julio-Claudian Rome* (Princeton: Princeton University Press, 2001), 234.

4 Quintus Curtius, *History of the Wars of Alexander* 8.6-8; trans. John Digby Esq., (London: A. Millar, 1747), Archive.org, accessed July 28, 2021. https://archive.org/details/quintuscurtiush00freigoog.

5 Roller, *Constructing Autocracy*, 236.

6 Seneca, *On Mercy* 1.17-1.18.

7 Ibid., 1.11-1.12.

8 *Diritas* is the Latin word for cruelty; Kenneth Scott, "The *Diritas* of Tiberius," *The American Journal of Philology* 53, no. 2 (1932): 151.

9 Ibid., 139.

10 Marcel Le Glay, et al., *A History of Rome,* 4th ed. (Malden: Wiley-Blackwell Publishing Company, 2009), 259.

11 C. Suetonius Tranquillus, *The Lives of the Twelve Caesars* 3.37; trans. Alexander Thomson, M.D. (London: George Bell & Sons, 1893), Archive.org, accessed July 31, 2021. https://archive.org/details/livestwelvecaes00foregoog/page/n228/mode/2up.

12 Ibid., 34.1.

13 Kenneth Scott, "Tiberius's Refusal of the Title "Augustus," *Classical Philology* 27, no. 1 (1932): 44.

14 Dio, *Roman History* 57.8.3.

15 Seneca, *On Mercy* 1.10.

16 David Shotter, *Tiberius Caesar*, 2nd ed. (New York: Routledge/Taylor & Francis Group, 2004), 10.

17 Agrippa was one of the men Augustus had initially appointed for succession.

18 Matthew Bunson, *A Dictionary of the Roman Empire* (New York: Oxford University Press, 1995), 47.

19 Le Glay et al., *A History of Rome*, 255-56.

20 Cassius Dio, *Roman History* VII 57.19.1; trans. Ernest Cary, PhD (New York: Harvard University Press, 1924), Archive.org, accessed July 31, 2021. https://archive.org/details/L175CassiusDioCocceianusRomanHistoryVII5660/page/n171/mode/2up?q=maiestas

21 Cornelius Tacitus, *Annals* 3.4.1; The Perseus Catalog, accessed August 3, 2021. http://www.perseus.tufts.edu/hopper/text?doc=Perseus%3Atext%3A1999.02.0078%3Abook%3D3%3Achapter%3D4

22 Ibid., 2.83.3

23 Dio, *Roman History* 57.13.6.

24 Ibid., 57.19.1.

25 M. Cary and H. H. Scullard, *A History of Rome*. 3rd ed. (London: Palgrave Macmillan, 1975), 353-354.

26 Suetonius, *Life of Tiberius* 59.1.

27 Le Glay et al., *A History of Rome*, 259.

28 M. K. Thornton et al., "The Financial Crisis of A.D. 33: A Keynesian Depression?" *The Journal of Ecnonomic History* 50, no. 3 (1990): 659.

29 Suetonius, *Life of Tiberius* 75.1.

30 Tacitus, *Annals* 6.50.

31 Seneca's efforts to steer Nero in the right direction failed. In 65 AD Seneca was embroiled in a plot to kill the young emperor—a scheme known as the *Pisonian* conspiracy. While there was no proof of Seneca's involvement, Nero ordered Seneca to commit suicide. Seneca followed tradition by severing his veins and bleeding to death. Three years later, in 68 AD, emperor Nero's tax policies made him an enemy of the state. He fled Rome and had his secretary, Epaphroditos, help him commit suicide for fear of facing execution by the Senate.

Book Review: Robert L. O'Connell's *The Ghosts of Cannae: Hannibal and the Darkest Hours of the Roman Republic*

Michael G. Stroud
American Military University

O'Connell, Robert L. *The Ghosts of Cannae: Hannibal and the Darkest Hours of the Roman Republic.* New York, NY: Random House, 2010, 310 pp. ISBN: 978-1-400067022.

The tactical genius of Hannibal Barca (247–182 BC), sworn enemy of Rome, would bring all his martial gifts to bear at Cannae and the result would be near destruction for the Roman Republic, as by day's end, nearly 70,000 Romans, including one consul would be dead. So great was the destruction that "if it [were] possible to conceive of hell on earth, this human abattoir at Cannae must have been the equal of any hell that history in all its perversity has managed to concoct" (160). The epochal catastrophe of Cannae, its buildup of preceding events, and subsequent aftershocks all make the case for Hannibal winning the battle but losing the war in Robert O'Connell's *The Ghosts of Cannae*. The author builds a compelling narrative that proposes that Cannae was more than just the destruction of a massive Roman army by its most vehement enemy, but that the effects or rather the ghosts of Cannae haunt military thinkers and strategists to this day.

O'Connell approaches the battle of Cannae by laying it out as if the entire affair were a dramatic play, a dramatic play with the deadliest of acts. He provides a cast of characters with such little-discussed personalities as Appius Claudius (Roman survivor of Cannae and compatriot of Scipio Africanus), supplemented with an appropriate number of easy-to-read maps, all in conjunction "acts" that build-up to and through the pivotal event of Cannae. O'Connell also ensures his readers are not lost in the linguistics of the day and has thoughtfully provided a very useful

 doi: 10.18278/sshj.10.3.8

five-page glossary of terms, that include key Latin, military, and technical terms.

The Ghosts of Cannae is strongly driven by the histories of Polybius and Livy. O'Connell does not take these historians at face value but leverages modern findings and scholarship to challenge their assertions and provide the reader a more modern, educated assessment of the subject matter, ranging from the uncertainty of the actual placement of the Aufidus River to the killing rate that would have been necessary in their destruction of the Roman army at Cannae within the historically given time frame. This is exemplified in the discussion of the elections of the Roman consuls Gaius Terentius Varro (fl. 3rd century BCE) and Lucius Aemilius Paullus (d. 216 BCE) who faced off against Hannibal. The author points out how both Polybius and Livy, who were pro-Roman, went to great lengths to lay the blame of the disaster of Cannae at the feet of Varro, while presenting Paullus as a fallen patriot. O'Connell cautions the reader to be wary of this claim as post-Cannae "the apparently incompetent Varro was given a number of other important commissions and even military commands" (134).

O'Connell utilizes visceral and evocative language in his narrative of the brutality and horror of the battles that led up to and including Cannae. This imagery, while not overblown, creates a sense of militaristic realism for the reader, which serves to both showcase the violence of the battles and the human toll that it took on all sides. *The Ghosts of Cannae* showcases this in the description of the final destruction of the encircled Roman army when the "systemic butchery [of the Romans continued] until the sunset on this terrible day" (157). The author builds upon this sensory engagement by going explaining the work of other sources (such as Victor Davis Hanson and Adrian Goldsworthy) of how many Romans would have had to have been killed every minute over eight hours to have accounted for over sixty-five thousand Romans killed. More recent scholars such as Yozan Mosig at the University of Nebraska-Kearney have taken to exploring the psychology of Hannibal and his war against Rome, including the refuting of the oft told tale of a young Hannibal and his swearing of eternal hatred of Rome.

The linear narrative weaves together disparate pieces of information from both primary and secondary sources in a masterful way that informs the reader while avoiding the pitfalls of previous knowledge assumption. *The Ghosts of Cannae* provides a strong tapestry of the entirety of the Hannibalic War (more commonly known as the Second Punic War) of 218–201 BC, from its root causes, the apex event of Cannae and its seismic ramifications (such as the eventual complete destruction of Carthage in 146 BC and the rise of the Roman Empire in 27 BC). This is accomplished by O'Connell's intuitive ability to concisely leverage the most informational and narratively important pieces of historic sources in balance with modern secondary sources, material, and scholarship to detail the military (the first double-envelopment in military history), political and personal significance of the entire campaign.

The Ghosts of Cannae is deeper and more engaging than just another assessment of Rome's greatest military disaster. The work passionately puts on display a narrative of two Mediterranean behemoths: the African maritime power of Carthage under the brilliant leadership of Hannibal Barca and that of the expansionist and growing Roman Republic. The personalities and brutality of the era are boldly presented giving life to them, their battles, and their consequences. The biggest deficit for O'Connell is the lack of source material from the Carthaginian perspective and therefore, a more Roman nuanced narrative. Even with this challenge, the author executes a carefully balanced telling of the entirety of the battle as well as the tragedy and eventual redemption of its Roman survivors.

The Ghosts of Cannae will educate the laymen while providing depth and understanding to the student of history. This engrossing work is a must for all who are looking for an engaging, gritty dive into the totality of history's most decisive battle and its aftermath, forever known as Cannae.

Book Review: Mary Beth Norton's *1774: The Long Year of the Revolution*

Gerald J. Krieger
Salve Regina University

Norton, Mary Beth. *1774: The Long Year of the Revolution.* New York: Alfred A. Knopf, 2020. 528 pp. ISBN: 9780385353366.

Although coffee was introduced to America by Captain John Smith with the founding of the Colony of Virginia in 1607, the beverage did not become common until the famous Boston Tea Party (December 16, 1773), when a group of merchants and traders disguised as Native Americans boarded the *Dartmouth*, *Eleanor*, and *Beaver* and threw 342 chests of tea into the Boston Harbor, valued at £18,000 (roughly the equivalent to one million U.S. dollars today), to oppose the tea tax. Only then did the colonists turn to coffee as a replacement for their favorite caffeinated beverage. Many historians are specializing in American history focus on the revolutionaries and the series of events that began with the Stamp Act riots (1765), culminating in the Boston Tea Party. Historians often suggest that this sequence of events "all seem to lead inexorably, if not almost inevitably, to the revolutionary war that followed," according to Mary Beth Norton (xvi). However, the events that followed December 1773 revealed many debates, disagreements, and disruptions rather than a unified voice of the colonies. Norton's latest book, *1774: The Long Year of the Revolution*, examines sixteen crucial months after the Boston Tea Party.

Norton dispels the myth that the Bostonians were the leaders of the rebellion, proving through a detailed examination of custom records, "Bostonians were somewhat more law-abiding than Philadelphians or New Yorkers. The imported more than 265,000 pounds of duty-paying tea in 1771 and over 107,000 the following year" (7). In fact,

doi: 10.18278/sshj.10.3.9

after 1770, East India Company (EIC) records demonstrate that Boston residents abandoned the boycott and began paying tea duties, much to the chagrin and criticism from other colonists (9). Tea was a very popular item at the time. Colonists in North America consumed over 265,000 pounds of tea in 1771 and 575,000 pounds of smuggled tea (7). Norton's study provides fresh insight into the period because she focuses on the written record of pamphlets, newspapers, and personal correspondences to underscore that very early it was the Philadelphians who led the opposition, rather than Samuel Adams and the Son of Liberty. For example, Norton provides compelling evidence that the year 1774 must be considered just as crucial to American independence as the pivotal years of 1775 and 1776 as a host of primary documents such as meeting minutes, pamphlets, letters, and newspaper articles highlight the fact that the public was leaning heavily towards independence long before the first shots were fired at the Battles of Lexington and Concord on April 19, 1775.

Norton argues that most historians gloss over 1774, focusing more on Lexington and Concord to explain the colonial frustrations leading up to the rebellion. Her detailed account relating events week by week provides details that most historians typically miss. Most narratives only cover six ships loaded with tea bound for the colonies, though there was a seventh, the *William*, which sank off the coast of Cape Cod in the middle of December (39). The author does a thorough job of piecing together the incomplete record of the shipwreck of the *William* while explaining how the salvaged tea was distributed. Ultimately, the colonists used the *Eunice*, a Salem fishing schooner, to move the fifty-four chests and one tea barrel to Castle William.

On November 28th, the first ship, the *Dartmouth* (Captain James Hall), sailed into Boston Harbor, followed by the *Eleanor* on December 3rd and the *Beaver* on December 7th (21). This event distorted the narrative and made Boston the center of the tea controversy when the center was really New York and Philadelphia. It is ironic that of the six ships that sailed to the colonies, the two cities that took the lead in the initial opposition to the tea did not have to address the physical presence of the tea because the *Polly*, under Captain Samuel Ayres, reached the Delaware River on December 25th, while the *Nancy* under Captain Benjamin Lockyer arrived in New York harbor in the middle of April 1774 (22). Another unique twist is that the Tea Act was enacted principally to assist the failing East India Company rather than as an attempt to raise revenue or coerce the colonists. The act *lowered* the duty on tea by cutting out the London middlemen.

Norton's story of the destruction of the *Peggy Stewart*, owned by Anthony Stewart and his father-in-law, which arrived in Annapolis on October 14th, captures the tension and confusion between those colonists enforcing the ban on tea and those attempting to comply (203). She takes the reader step by step as Stewart ordered British goods, hoping to beat the November 1st deadline for nonimportation outlined in the June Virginia Association. Stewart was transporting orders for several customers. Among the many products on the ship was over two thousand pounds of EIC tea (203). Despite the ship's arrival two weeks before the ban, the local committee required that the tea be turned over to the committee, which they did, along with offering to pay all of the duties required other than that on tea. Still, the customs inspector refused and demanded payment for the entire shipment. The duties were subsequently paid, prompting a local committee meeting when they learned that EIC tea was among the contents (203). The committee elected to bring the tea onshore and burn it, though the group decided an apology was also required. Others on the committee began to demand that the ship be burned. Others insisted that Stewarts' house be burned as well (his pregnant wife was resting inside). Eventually, Stewart, and the Williams brothers who had ordered the tea, were forced to read a statement aloud, admitting they had insulted the liberties of America. After this admission, Stewart was forced to burn his ship down to the waterline, which he did fearful of "the Fury of a lawless Mob," as he later admitted (204).

Norton's detailed analysis fills in the gaps and details week by week as events unfolded in printed media around the colonies for the sixteen months that followed the Boston Tea Party. Her book fills a crucial gap during this period. It informs general readers and historians alike who seek to understand the sequence of events that propelled the colonists on a collision course with London during the early years of the American Republic.

Saber and Scroll Exhibition Review Contest - First Prize

Pexcho's American Dime Museum

Aisha Manus
American Military University

In an unassuming building at 216 6th St., in Augusta, Georgia, a collection of extraordinary wonders lives, begging to be explored. From shrunken heads to a vial of flatulence, Victorian medical devices to a live sword swallower, Pexcho's American Dime Museum is a step back to a time of American curiosity and gullibility. Is Pexcho's a real museum? You bet! Is it also a possible scam? Probably! What is real and what is fake is left to the observer to decide, and that is just part of the fun!

Museum of the Strange and Weird Open! reads a colorful sign just inside the door. A caricature of a gangrenous hand, missing some flesh down to the bone, points to the entrance of the theater. Photo by Aisha Manus.

 doi: 10.18278/sshj.10.3.10

Dime museums were extremely popular attractions during the Progressive Era, 1890-1920, as a cheap way to entertain the laborers of major metropolitan cities, but their history began in the early 1800s. The most notable dime museums were owned by P.T. Barnum in New York in the 1840s, where he turned the museums into what they are known for—educational entertainment. This same type of educational entertainment is a constant at the American Dime Museum, as curator Peter Excho personally guides visitors through a two-hour tour where he not only gives a detailed history of dime museums and how they work, but also of his amazing exhibits on display.

The tour begins in his 127-seat theater, where Peter first details the initial history of Dime museums, focusing most of his stories on Barnum and his endeavors, as he is the father of zoos and entertainment. During this history lesson, Peter engages with the crowd, in particular the children, asking questions and inspiring curiosity. He consistently reminds the crowd that the children in the audience that day were probably more educated than the average dime museum attendee at the time, so it makes sense the only real thing guests of the past learned in a dime museum was how to lose their wallets and little else! Dime museums are, after all, where the phrase "to nickel and dime someone" comes from! After about 15 minutes of interesting lectures, he declares to the audience, "Let's go look at some real fake stuff!"

Only about 40 percent of his collection is on display at the moment as the building is not very large, but it is still an impressive collection of the rare, unique, and fake oddities of the world. An avid collector of oddities his whole life, he acquired some of his collection from the former Baltimore Dime Museum after it closed, while several of the pieces in his collection are on loan or donated by enthusiastic patrons. One particular patron of his museum donated an African wedding ring, which he claims only two other museums in the world, the Brooklyn and the Metropolitan, have in their collections—but only his is on display. He displays this ring that comes from the African Ivory Coast with toys from the Asian island of Borneo. Keeping with the theme of rare items most people would never get to see, next to this display are rare cubular spotted eggs of a bird found only in the Nevada desert, which I must confess were an exciting sight to see in person.

While we are in this first small room of the museum, Peter continues to give us a history of the dime museums of the past and in turn displays a collection of cabinet cards, which were cards given by performers people paid to see, such as the bearded lady or world's fattest man. He also has an original slapstick on display, as the origin of slapstick comedy is in the dime museum, as is the origin of vaudeville. In this room, guests are encouraged to touch the real foot of a rhinoceros, and nearly everyone does! We then move onto the next small area which is a collection of local Augusta artifacts, such as elixir and snake oil bottles. Other delicate items, such as Victorian-era cat-shaped toys made with real cat teeth and a tie

Here curator Peter Excho educates the group on the history of dime museums in America from the stage of his 127-seat theater. The hall also serves as a venue for resident performers, as well as for traveling acts from time to time, and sword swallower "Red." Photo by Aisha Manus.

made from the feathers of a pheasant decorate this room as well.

At this point, it appears as if the museum tour has ended as there is nowhere else to go until he opens a false wall, and we enter the part of the museum that is a little more grotesque but still incredibly kid-friendly. He has an original breast pump, where he gives the history of how they often killed babies due to the build-up of bacteria. There are also key chains and watch bands made from real human hair! Then he brings out a vintage electrode device and encourages his guests to shock themselves. Only one guest during my visit was brave enough to do so, only to discover that while it may have looked and sounded scary, there was very little pain. He explains that these devices, while completely ineffective, are still in use to this day because medical offices, usually medical spas or dermatological offices, use words like "might" or "could" help with complaints, proving that people still can be duped out of their money just as they were in the nineteenth century?

The next section is the most macabre of his collection, where he has on display his most grotesque items. It is in this room where were learn that the

Guinness World Record holder John “Red” Stuart performing one of the seven deadly tortures of sword swallowing before an awestruck audience. Instead of his regular sword, Red is swallowing a 9-pound axel from a Ford Model T. Photo by Mr. Excho’s eldest son.

Filed under "Weird" human hair trinkets from the Civil War era are on display. What better way to think of your loved one than to wear a bracelet made from their hair? So romantic. Photo by Aisha Manus.

A brave guest smelling the rancid, putrid smell of "Death." The pungent odor is just as foul in the wintertime as it was in the summer, so guests year-round get the full experience. Photo by Aisha Manus.

body of famous American Outlaw Jesse James was moved around the U.S. to various dime museums to profit off the curiosities of people and our fascination with the dead. While he does not have a body in this room, he does have several heads! From regular human heads in jars to shrunken heads, these heads could be real, or they could fake—we don't know! He does, however, teach us how fake shrunken heads were created using goatskin to meet the demands for them during the height of their popularity; he never clarifies which kind he has! The most interesting shrunken head in his collection is fashioned into a musical instrument, allowing a person to play guitar with the dead! It is at this point in the tour when Peter then offers to turn you into a shrunken head when you die if you wish, letting his guests know that if they will him their body, he will happily provide the service to enact your last wishes. His own ex-wife is said to have willed her skeleton to the museum upon her death, requesting that she be displayed with a cigarette in hand. Though it might be some years until his ex-wife's skeleton is placed in the museum, Peter informs us that he is currently working on having the skeleton of a T-rex brought to the museum, thanks to one of his patrons. What makes this extra delightful is the fact that all dinosaur bones on display in any museum are always fake, so these fake bones would fit right in with this museum of real fake things!

We move onto the next section of the museum where we were treated with a glimpse of a living pigmy hippo brought in by one of his children, but it was feeding time, so the hippo did not stay long. But that is a good thing as I do not think the hippo would have enjoyed the next portion of the tour, when Peter invited his guests to smell the scent of death. Having attended his museum in both summer and winter, even in the cold the smell is a putrid, rancid concoction only the brave dare smell. You've been warned! Interestingly, it is in this room, however, where Peter claims there is a body! Inside the pauper's coffin displayed in the room lies the body of a man (so we are told). Found on the street dead and disfigured in his features to make people consider him an oddity, the man was preserved and displayed for many years until he finally came to the American Dime Museum. Wanting to give the man a proper burial, Peters says that he is waiting for the world to truly open up again so that he can have the entertaining funeral that the man deserves, since "the first word found in funeral is fun, after all!" I eagerly await my invite!

Throughout the entire tour, upbeat ragtime piano music plays in the background, giving the tour just the right touch of additional entertainment while keeping to the allusion that one has stepped back in time. This is especially entertaining when he shows us his final oddity in his collection. In one vial, he claims the contents are the flatulence of Joesph Pujol, a famous French stage performer by the name of Le Pétomane, whose ability to control his abdominal and rectal muscles enabled him to fart on command and make multiple tones.

The front of Pexcho's American Dime Museum. The outside may be unassuming but don't judge a book by its cover for the inside is full of wonders and amazement. Photo by Aisha Manus

So gifted in his abilities to make music and other sounds with his gas, Le Pétomane performed for kings! To know his fart has been contained in that vial for at least a hundred years and sits just a few feet from me is truly magnificent.

After walking around for about an hour and a half the tour of his collection finishes by walking through a coffin and back into the theater where the audience is treated to a real-life "working act." John Stuart, who performs under the stage name "Red" for his once flaming red hair, and holds nine Guinness World Records in sword swallowing, gives the audience a joke-filled history lesson on the art of swallowing swords and the other seven deadly tortures performed at these types of museums of the past. He then demonstrates a few of the tortures. First, he does the famous "block head" trick, in which he hammers a long nail directly into the center of his head. He then demonstrates his sword-swallowing abilities. Red may be the oldest active sword swallower, but has no trouble as he individually swallows a sword, a glowing lightsaber, a wire coat hanger, and a 9-pound axel from a Ford Model T. I will admit, reader, that I did not actually watch him perform his tricks as I was not keen to pass out, but the six other guests that were there with me seemed delighted by his amazing skills.

The cost of the museum is a mere $14.28 for adults and just a dime for children nine and under; tours are currently by appointment only. The tour is wheelchair and handicapped accessible, and there are plenty of places to sit throughout, though a few places were a little difficult to maneuver. The lighting in the museum is low, to simulate the atmosphere of gas and oil lighting used in the past, so I consider it sensory-friendly for guests who need lower lighting. You can find more information about the museum on their Facebook page, https://www.facebook.com/americandimemuseum, or by calling (225) 448-1453 or emailing Peter at pexcho66@hotmail.com.

Call for Papers - Summer 2022

The Saber and Scroll Journal is currently soliciting articles, books, and exhibition reviews for its Summer 2022 edition.

This is an open edition and will consist of works on all topics of history and historiography. All historical time periods and geographic regions are welcome provided they address a topic of historical interest.

Book reviews and exhibition reviews should be on recent events or publications.

Students, alumni, faculty, and independent historians, from any and all academic institutions, are welcome to submit their original, previously submitted, and corrected coursework, provided it has not been published in another publication.

All submissions will be reviewed by a group of graduate students, alumni, and APUS faculty editors.

Furthermore, all submissions must adhere to the Saber and Scroll Journal's submission guidelines which can be located at:

https://saberandscroll.scholasticahq.com/for-authors.

The Saber and Scroll reserve the right to right to reject, without further review, any submission that does not follow these guidelines or meet our high academic standards.

Any questions should be directed to Jeff Ballard, Editor-in-Chief, Saber and Scroll Journal at: saberandscrolljournal@gmail.com.

Abstract (124 words max) Deadline: May 1, 2022

Manuscript Deadline: June 1, 2022

Saber & Scroll Historical Journal

SUBMISSION GUIDELINES • MARCH 1, 2022

To assist our Editorial Staff in processing your manuscript efficiently, please follow these basic guidelines. Taking these steps minimizes work duplication on your part to ensure prompt publication. We reserve the right to reject, without further review, any submission that does not meet the minimum guidelines or meet our high academic standards.

—The Editorial Staff

1. Historically, the Journal is published 3-4 times a year; Spring (March), Summer (June), Fall (September), and Winter (December). Only Winter is a themed issue, lately dedicated to military history.

2. Feature articles should be between 2,500 and 8,000 words in length, not including front or back-matter. Features longer than the maximum length may be split into two or more parts (serialized) for publication in consecutive issues.

3. Book reviews should be between 800 and 1200 words. An image (*.jpg) of the book cover should be attached to the submission email. Do not embed in the document. Book reviews are in high demand, so they receive immediate attention and are fast-tracked for publication.

4. Submissions of museum and exhibition reviews are highly prized by the Journal, which periodically sponsors contests, with non-cash prizes, for their publication. Please contact the Editor-in-Chief for any contest rules and the special submission guidelines.

5. Only manuscripts submitted in Microsoft Word (*.doc), and images in *.jpg and *.heic formats will be accepted. No *.pdf or *.tiff files.

6. The Journal's citation style is Chicago-Turabian, using endnotes, numbered in Arabic, super-script numerals, and including a complete, correctly formatted bibliography.

7. The manuscript narrative and notes should be in 12-point Times New Roman, and the paragraphs double-spaced with a one-inch margin on all sides of the page.

8. A cover page, separate from the manuscript, must contain the following infor-

cont'd.

mation: author's name as they would like it to appear in print, school affiliation (if any), email address, and phone number.

9. The cover sheet must also include a statement affirming the manuscript has not been previously published or submitted elsewhere for publication.

10. The author's name should not appear on the manuscript to preserve the anonymity of the double-blind review process.

11. In addition to the narrative, each feature article must include a 100-150 word abstract and a list of 8-10 keywords.

12. We encourage authors to incorporate at least three (but no more than five) images into their feature articles and museum reviews. Images can be photographs, paintings, illustrations, or maps. However, images must be of the author's creation or in the public domain.

13. Each image should be attached to your submission email as a separate file, not embedded in the manuscript. Label the image in the order in which you would like them to appear in your published article, e.g., "Image 1," "Image 2," etc.

14. In a single Word document provide a caption and source information for every image beside the file name you gave to each. Ex: "Picture 1 - PhineasTaylor Barnum, July 5, 1810 – April 7, 1891, photographed by Matthew Brady. National Portrait Gallery NPG.97.63."

15. Our publisher prints the Journal using black text and grayscale images. Images should be approximately 300 dpi resolution. However, the online version of the Saber and Scroll is published in full color so, whenever possible, select color images.

16. The Journal reserves the right to publish manuscripts without any of the author-supplied images, substitute images that do not meet the minimum standards, or limit the number of images in an article or an issue because of space limitations.

17. Every issue of the Saber and Scroll published since its inception in 2013 can be viewed in its entirety at https://saberandscroll.scholasticahq.com/. Prospective authors are encouraged to browse the online issues and read the submission guidelines several times before submitting a manuscript for consideration.

18. Every published author will receive a free copy of the issue in which their manuscript appears, mailed to their home compliments of our publisher. Copies of the current and select past issue of the Journal may be purchased at Amazon.com.

19. Finally, the Journal is always looking for volunteers to assist with various tasks, especially those related to quality control of our new web portal and uploading images for older issues. Volunteers will receive a year's subscription to the Journal (3-4 issues) as our thanks for your participation.

To request additional information or submit proposed manuscripts for consideration, please email the Editor-in-Chief at

saberandscrolljournal@gmail.com.

Bunker Diplomacy: An Arab-American in the U.S. Foreign Service
by Nabeel Khoury

After twenty-five years in the Foreign Service, Dr. Nabeel A. Khoury retired from the U.S. Department of State in 2013 with the rank of Minister Counselor. In his last overseas posting, Khoury served as deputy chief of mission at the U.S. embassy in Yemen (2004-2007).

Managing Challenges for the Flint Water Crisis
Edited by Toyna E. Thornton, Andrew D. Williams, Katherine M. Simon, Jennifer F. Sklarew

This edited volume examines several public management and intergovernmental failures, with particular attention on social, political, and financial impacts. Understanding disaster meaning, even causality, is essential to the problem-solving process.

The Forgotten Army: The American Eighth Army in the Southern Philippines 1945
by Robert M. Young

History has produced many famous armies. It has also produced several that few knew even existed. The American Eighth Army of World War II is one such force. They saw action throughout the Southwest Pacific, specifically in the Philippines.

Issues in Maritime Cyber Security
Editors: Dr. Joe DiRenzo III, Dr. Nicole K. Drumhiller, Dr. Fred S. Roberts

The complexity of making MTS safe from cyber attack is daunting and the need for all stakeholders in both government (at all levels) and private industry to be involved in cyber security is more significant than ever as the use of the MTS continues to grow.

Contests of Initiative: Countering China's Gray Zone Strategy in the East and South China Seas
by Dr. Raymond Kuo

China is engaged in a widespread assertion of sovereignty in the South and East China Seas. It employs a "gray zone" strategy: using coercive but sub-conventional military power to drive off challengers and prevent escalation, while simultaneously seizing territory and asserting maritime control.

This publication is available open access at:
http://www.ipsonet.org/publications/open-access

Thanks to the generosity of the American Public University System

APU American Public University

AMU American Military University

Made in United States
North Haven, CT
05 November 2026

21772463R00085